Courtesy of Bryn Mawr College

Emmy Noether in Bryn Mawr

Proceedings of a Symposium Sponsored by the Association for Women in Mathematics in Honor of Emmy Noether's 100th Birthday

Edited by
Bhama Srinivasan and Judith D. Sally

With Contributions by

Armand Borel Walter Feit Nathan Jacobson Jeanne LaDuke
Marguerite Lehr Ruth S. McKee Uta C. Merzbach
Emiliana P. Noether Gottfried E. Noether Grace S. Quinn
Judith D. Sally Richard G. Swan Olga Taussky
Karen Uhlenbeck Michele Vergne

Springer-Verlag
New York Berlin Heidelberg Tokyo

Bhama Srinivasan
Department of Mathematics
University of Illinois
Box 4348
College of Liberal Arts
and Sciences
Chicago, IL 60680
U.S.A.

Judith D. Sally
Department of Mathematics
Northwestern University
College of Arts and Sciences
Lunt Hall
Evanston, IL 60201
U.S.A.

AMS Subject Classifications: 01-06, 01-60, 12-06

Library of Congress Cataloging in Publication Data
Main entry under title:
Emmy Noether in Bryn Mawr.
Bibliography: p.
Contents: Brauer factor sets, Noether factor sets and crossed products / N. Jacobson— Noether's problem in Galois theory / R. G. Swan—Noether normalization / J. D. Sally—[etc.]
1. Mathematics—Congresses. 2. Noether, Emmy, 1882–1935. 3. Mathematicians—United States—Biography. I. Noether, Emmy, 1882–1935. II. Sally, Judith D. III. Srinivasan, Bhama, 1935– . IV. Association for Women in Mathematics (U.S.)
QA1.E52 1983 510 83-651

With 16 illustrations.

Softcover reprint of the hardcover 1st edition 1983

Typeset by Composition House Ltd., Salisbury, England.

9 8 7 6 5 4 3 2 1

ISBN-13:978-1-4612-5549-9 e-ISBN-13:978-1-4612-5547-5
DOI: 10.1007/978-1-4612-5547-5

Preface

This volume contains the proceedings of a Symposium held in honor of Emmy Noether's 100th birthday which was sponsored by the Association for Women in Mathematics, and held at Bryn Mawr College on March 17, 18 and 19, 1982. It was fitting that the Symposium be held at Bryn Mawr, where Noether held her last position. Indeed, the lectures were held in Goodhart Hall, where the famous Memorial Address was delivered by Hermann Weyl on April 29, 1935. The Association for Women in Mathematics is honored to have sponsored this event, which was judged by many of those attending to have been not only scientifically successful but a specially moving occasion.

There were nine scientific lectures by Nathan Jacobson, Richard Swan, Judith Sally, David Mumford, Michele Vergne, Olga Taussky-Todd, Karen Uhlenbeck, Walter Feit, and Armand Borel. There was also a panel discussion on "Emmy Noether in Erlangen, Göttingen, and Bryn Mawr" in which Gottfried Noether, Olga Taussky-Todd, Grace Quinn, Ruth McKee, and Marguerite Lehr participated. The last four were at Bryn Mawr during Emmy Noether's time and presented their personal reminiscences of her. Gottfried Noether is a nephew of Emmy Noether and gave an account of her life and career in Germany.

We present here articles by all the speakers (with the exception of David Mumford) based on their talks at the Symposium. The first speaker was Nathan Jacobson, who knew Emmy Noether personally at the Institute for Advanced Study and who has recently edited her *Collected Works*. Various areas in which Noether made her contributions were covered by the speakers as follows: Jacobson, crossed products; Sally, commutative ring theory; Swan, invariant theory and Galois theory; Mumford, invariant theory; Vergne, representation theory; Taussky-Todd, number theory; Uhlenbeck, differential invariants (and their applications to physics). It might be mentioned that these speakers also emphasized recent work in these and other areas, as did Feit who talked about finite group theory, and Borel who talked about L^2-cohomology and intersection cohomology.

We have also included in this volume the presentations of the speakers at the panel discussion. We hope that this will add a personal touch to the volume. Finally, we include articles by Jeanne LaDuke and Uta C. Merzbach which are based on talks given at a Symposium sponsored by the Association for Women in Mathematics at the annual meeting of the American Mathematical Society in Cincinnati in January, 1982. They deal with the development of algebra in the U.S.A. and Germany, respectively, prior to and during the time of Emmy Noether.

We take this opportunity to thank Bryn Mawr College for hosting the Symposium and the American Mathematical Society for their cooperation and generous support in the arrangements of the Symposium.

Chicago,
November 29, 1982

BHAMA SRINIVASAN
JUDITH D. SALLY

Contents

Brauer Factor Sets, Noether Factor Sets, and Crossed Products*

NATHAN JACOBSON†

The role of Noether's crossed products and factor sets in the study of the Brauer group Br(F) of a field F is well known. In particular, it is central in the determination of the Brauer group of a number field and in the proof of the Albert–Brauer–Hasse–Noether theorem that central division algebras over number fields are cyclic ([2], [5], [8], [9]). The central algebraic result of Noether's theory is the isomorphism of the subgroup Br(E/F) of Br(F) consisting of the algebra classes having a finite dimensional Galois extension field E/F as a splitting field with the cohomology group $H^2(G, E^*)$ where $G = \text{Gal } E/F$. This leads to an isomorphism (given later) of the full Brauer group Br(F) with a cohomology group of the Galois group of the separable algebraic closure of F.

On the other hand, the main algebraic results on Br(F) (e.g. the fact that this group is a torsion group) were first derived by Brauer using Brauer factor sets. It is interesting to note also that in what may have been Noether's first presentation of crossed products, namely, that given in her lectures in Göttingen in the summer semester of 1929, the crossed product theory was preceded by an account of Brauer's factor sets. The notes for these lectures were prepared by Deuring and will appear in [14]. Brauer factor sets were used by Weyl in his study of Riemann matrices and they are also implicit in Albert's original treatment of the subject ([1], [15]). A lesser known aspect of Brauer's theory is a general construction of central simple algebras from a separable extension field K/F and a Brauer factor set with non-zero values in the normal closure E/F of K ([3]). This construction was used by Brauer to show that if $\mathscr{A}$ is a central division algebra of degree 5 over F then there exists an extension field E/F obtained by successive adjunctions of three square roots and a cube root such that $\mathscr{A}_E$ is a cyclic algebra ([4]).

* This work was partially supported by the National Science Foundation Grant MCS79-05018.

† Department of Mathematics, Yale University, P.O. Box 2155, Yale Station, New Haven, CT 06520, U.S.A.

In this paper we give a new derivation and extension of the theory of Brauer factor sets. Our starting point is a special type of generation of any finite dimensional central simple algebra $\mathscr{A}$ over F of degree n by a commutative separable subalgebra $K = F[u]$ of dimension n and another element v. ($\mathscr{A}$ contains plenty of such subalgebras.) We show that there exists an element v such that $\mathscr{A} = KvK$. If E/F is a splitting field for K in the sense that E is a minimal extension field of F such that $K_E = Ee_1 \oplus Ee_2 \oplus \ldots \oplus Ee_n$ then E is a splitting field for $\mathscr{A}$. Hence we have an imbedding of $\mathscr{A}$ in the matrix algebra $M_n(E)$ so that $M_n(E) = E\mathscr{A}$ and $u = \operatorname{diag}\{r_1, r_2, \ldots, r_n\}$ where the r_i are distinct elements of E. Then v is a matrix (v_{ij}) all of whose entries are $\neq 0$ and $c = \{c_{ijk}\}$ where $c_{ijk} = v_{ij}v_{jk}v_{ik}^{-1}$ is a Brauer factor set. This leads to the general construction of central simple algebras due to Brauer.

Formulated in a more abstract fashion, we can define a Brauer factor set c in the following way. Let K be a finite dimensional commutative separable algebra over F, E/F its splitting field (unique up to isomorphism) and let M be the set of homomorphisms of K/F into E/F. Then $|M| = [K:F]$. A Brauer factor set is a map of $M \times M \times M$ into E^* that is homogeneous in the sense that for any $\sigma \in \operatorname{Gal} E/F$, $c(\sigma\alpha, \sigma\beta, \sigma\gamma) = \sigma c(\alpha, \beta, \gamma)$, $\alpha, \beta, \gamma \in M$ and

$$c(\alpha, \beta, \gamma)c(\alpha, \gamma, \delta) = c(\alpha, \beta, \delta)c(\beta, \gamma, \delta).$$

Let $\mathscr{B}(K, c)$ denote the set of homogeneous maps l of $M \times M$ into E. These constitute a vector space over F and we can define a product in $\mathscr{B}(K, c)$ by

$$ll'(\alpha, \beta) = \sum_{\gamma \in M} l(\alpha, \gamma)c(\alpha, \gamma, \beta)l'(\gamma, \beta).$$

Then $\mathscr{B}(K, c)$ is central simple and every finite dimensional central simple algebra over F can be obtained in this way.

The Brauer factor sets form an abelian group under multiplication of images in E^*. This contains the subgroup of elements of the form

$$(\alpha, \beta, \gamma) \rightsquigarrow l(\alpha, \beta)l(\beta, \gamma)l(\alpha, \gamma)^{-1}$$

where l is a homogeneous map of $M \times M$ into E^*. We can form the factor group which we denote as $H^2(K/F)$. On the other hand, the subset $\operatorname{Br}(K/F)$ of the Brauer group $\operatorname{Br}(F)$ of the classes of central simple algebras over F that contain an $\mathscr{A}$ of degree n containing K is a subgroup of $\operatorname{Br}(F)$ and $\operatorname{Br}(K/F) \cong H^2(K/F)$. If K is a field then $\operatorname{Br}(K/F)$ is the kernel of the homomorphism of $\operatorname{Br}(F)$ into $\operatorname{Br}(K)$ defined by $\{\mathscr{A}\} \rightsquigarrow \{\mathscr{A}_K\}$.

The Noether theory can be obtained by specializing to the case in which $K = E$. In this case if $\mathscr{A}$ is a central simple algebra of degree n containing the Galois extension E/F of dimension n then $\mathscr{A}$ is a crossed product (E, G, k) of E with its Galois group G and its Noether factor set k. On the other hand, we know also that $\mathscr{A} = \mathscr{B}(E, c)$ for a Brauer factor set c. It is easy to establish the relations between the two constructions and between the Brauer factor sets and the Noether factor sets. In this way we obtain Noether's results, in particular, the basic isomorphism of $\operatorname{Br}(E/F)$ with the second cohomology group $H^2(G, E^*)$.

In the last section we give an application to the theory of central simple algebras with involution. If the Brauer factor set c is symmetric in the sense that $c(\alpha, \beta, \gamma) =$

$c(\gamma, \beta, \alpha)$ for $\alpha, \beta, \gamma \in M$ then $l \rightsquigarrow {}^t l$ where ${}^t l(\alpha, \beta) = l(\beta, \alpha)$ is an involution of orthogonal type in $\mathscr{B}(K, c)$. Moreover, we can show that any central simple algebra with an involution of orthogonal type is isomorphic as algebra with involution to a pair $(\mathscr{B}(K, c), t)$.

1. Reduced Characteristic Polynomial, Trace and Norm

The concept of the reduced characteristic (or generic or minimum) polynomial of a finite dimensional associative algebra has been defined by the author in [10]. An extension of the theory to strictly power associative algebras is given in [12], where many properties of the reduced characteristic polynomial are derived. In this section we shall derive some properties and applications that will be required.

Let $\mathscr{A}$ be a finite dimensional associative algebra over a field F, $(u_1, u_2, \ldots, u_n)$ a base for A/F. If $a \in \mathscr{A}$ we denote the minimum polynomial of a in $\mathscr{A}$ by $\mu_a(\lambda)$. We recall some well-known results on $\mu_a(\lambda)$ in the special case in which $\mathscr{A} = M_m(F)$ (so $n = m^2$). In this case we have the characteristic polynomial $\chi_a(\lambda) = \det(\lambda 1 - a)$ and the Hamilton–Cayley theorem that $\chi_a(a) = 0$. It follows that $\mu_a(\lambda) | \chi_a(\lambda)$. We recall also that we can diagonalize the matrix $\lambda 1 - a$ in $M_m(F[\lambda])$, that is, we can find invertible matrices $P(\lambda)$ and $Q(\lambda)$ in $M_m(F[\lambda])$ such that

$$P(\lambda)(\lambda 1 - a)Q(\lambda) = \operatorname{diag}\{d_1(\lambda), d_2(\lambda), \ldots, d_m(\lambda)\}, \tag{1.1}$$

where the $d_i(\lambda)$ are monic polynomials and $d_i(\lambda) | d_j(\lambda)$ if $i \leq j$. Then $\chi_a(\lambda) = \prod_1^m d_i(\lambda)$ and we have the sharpening of the Hamilton–Cayley theorem due to Frobenius: $d_m(\lambda) = \mu_a(\lambda)$. Evidently this implies that $\mu_a(\lambda)$ and $\chi_a(\lambda)$ have the same irreducible factors in $F[\lambda]$ and the same roots in the algebraic closure $\bar{F}$ of F.

We consider $\mathscr{A}$ again with the base $(u_1, u_2, \ldots, u_n)$. We introduce n indeterminates $\xi_1, \xi_2, \ldots, \xi_n$ and the field $F(\xi) \equiv F(\xi_1, \xi_2, \ldots, \xi_n)$ of rational expressions in the ξ_i with coefficients in F. Consider the algebra $\mathscr{A}_{F(\xi)}$ obtained by extending the base field of $\mathscr{A}$ to $F(\xi)$. Put $x = \sum \xi_i \mu_i \in \tilde{\mathscr{A}} = \mathscr{A}_{F(\xi)}$. We call this a *generic element* of $\mathscr{A}$. We now denote the minimum polynomial of x by $m_x(\lambda)$ (rather than $\mu_x(\lambda)$) and write this as

$$m_x(\lambda) = \lambda^m - \tau_1(\xi)\lambda^{m-1} + \cdots + (-1)^m \tau_m(\xi), \tag{1.2}$$

where $\tau_i(\xi) \in F(\xi)$. We now have

1.3 Lemma. *$m_x(\lambda) \in F[\lambda, \xi] \equiv F[\lambda, \xi_1, \ldots, \xi_n]$. More precisely, $\tau_i(\xi)$ is a homogeneous polynomial of degree i in the ξ's.*

Proof. Let ρ be a faithful representation of $\mathscr{A}$ by matrices. For example, we can take ρ to be the regular representation. Now ρ has a unique extension to a faithful representation $\tilde{\rho}$ of $\tilde{\mathscr{A}}$ and $\tilde{\rho}(x) = \sum \xi_i \rho(u_i)$ so the entries of this matrix are homogeneous linear expressions in the ξ's. Hence the characteristic polynomial $\chi_{\tilde{\rho}(x)}(\lambda)$ has the form

$$\lambda^N - t_1(\xi)\lambda^{N-1} + \cdots + (-1)^N t_N(\xi), \tag{1.4}$$

where $t_i(\xi)$ is a homogeneous polynomial of degree i in the ξ's and N is the degree of the representation. Since $\tilde{\rho}$ is a monomorphism, $m_x(\lambda)$ is the minimum polynomial of $\tilde{\rho}(x)$ and hence $m_x(\lambda) \mid \chi_{\tilde{\rho}(x)}(\lambda)$. It follows from Gauss' lemma that $m_x(\lambda) \in F[\lambda, \xi]$. Moreover, since $\chi_{\tilde{\rho}(x)}(\lambda)$ is homogeneous of degree N in λ and the ξ's, $m_x(\lambda)$ is homogeneous of degree m in λ and the ξ's. It follows that $\tau_i(\xi)$ in (1.2) is homogeneous of degree i in the ξ's. □

We now write $x^j = \sum_{i=1}^n \mu_{ji}(\xi)u_i$, $j = 0, 1, 2, \ldots$ where it is clear that $\mu_{ji}(\xi)$ is a homogeneous polynomial of degree j in the ξ's. The equation $m_x(x) = 0$ is equivalent to n equations

$$\sum_{j=0}^{m} (-1)^j \tau_j(\xi)\mu_{m-j,i}(\xi) = 0, \tag{1.5}$$

where $1 \le i \le n$, $\tau_0 = 1$. Now let $a = \sum \alpha_i u_i$, $\alpha_i \in F$ and put

$$m_a(\lambda) = \lambda^m - \tau_1(a)\lambda^{m-1} + \cdots + (-1)^m \tau_m(a), \tag{1.6}$$

where $\tau_i(a) = \tau_i(\alpha) \in F$. Then we have the relations $\sum (-1)^j \tau_j(\alpha)\mu_{m-j,i}(\alpha) = 0$ obtained by specializing $\xi_k \rightsquigarrow \alpha_k$ in (1.5). These imply that

$$m_a(a) = 0. \tag{1.7}$$

Now $m_a(\lambda)$ is independent of the choice of the base $(u_1, u_2, \ldots, u_n)$. For, if $(v_1, v_2, \ldots, v_n)$ is a second base and $v_i = \sum \beta_{ij}u_j$, $(\beta_{ij}) \in \mathrm{GL}_n(F)$ then $\sum \xi_i v_i = \sum \xi'_j u_j$ where $\xi'_j = \sum \xi_i \beta_{ij}$ and $m_{\sum \xi_i v_i}(\lambda)$ is obtained by replacing ξ_j by ξ'_j, $1 \le j \le n$ in $m_x(\lambda)$. Then if $a = \sum \alpha_i v_i$, $a = \sum \alpha'_j u_j$ where $\alpha'_j = \sum \alpha_i \beta_{ij}$. It follows that $m_a(\lambda)$ is unaltered in passing from the base $(u_1, \ldots, u_n)$ to the base $(v_1, \ldots, v_n)$. We shall call $m_a(\lambda)$ the *reduced characteristic polynomial* of the element a and

$$t(a) = \tau_1(a), \qquad n(a) = \tau_m(a) \tag{1.8}$$

the *reduced trace* and *reduced norm* respectively of a. The integer m is called the *degree of the algebra* $\mathscr{A}$.

Let E be an extension field of F and consider the algebra $\mathscr{A}_E$. We can regard $\mathscr{A}$ as contained in $\mathscr{A}_E$. Then the base $(u_1, \ldots, u_n)$ of $\mathscr{A}$ is also a base for $\mathscr{A}_E$. It follows that the reduced trace and reduced norm functions on $\mathscr{A}_E$ are extensions of these functions on $\mathscr{A}$. This remark implies that in developing the properties of these functions there is no loss in generality in assuming the base field is infinite or even algebraically closed. The advantage of dealing with infinite base fields is that we can use the Zariski topology of the vector space $\mathscr{A}/F$.

We define the *degree* of $a \in \mathscr{A}$ to be the degree of the minimum polynomial $\mu_a(\lambda)$ of a.

1.9 Proposition. *If F is infinite then the elements of $\mathscr{A}$ whose degrees are the degree of $\mathscr{A}$ constitute a non-vacuous Zariski open subset of $\mathscr{A}$.*

Proof. As before, we write

$$x^j = \sum_{i=1}^{n} \mu_{ji}(\xi)u_i, \qquad j \ge 0 \tag{1.10}$$

for the generic element x. Since the degree of the minimum polynomial $m_x(\lambda)$ of x in $\tilde{\mathscr{A}} = \mathscr{A}_{F(\xi)}$ is m, the elements x^k, $0 \leq k \leq m - 1$ are linearly independent in $\tilde{\mathscr{A}}$. Hence there exists a non-zero m-rowed minor in the $m \times n$ matrix $(\mu_{ji}(\xi))$. Let these non-zero m-rowed minors be $D_1(\xi), \ldots, D_q(\xi)$. Then it is clear that the set of elements $a = \sum \alpha_i u_i$ such that $\deg a = m$ is the union of the sets defined by $D_j(\alpha) \neq 0$. Evidently this is a non-vacuous open subset of $\mathscr{A}$. □

We again consider the faithful representation ρ of $\mathscr{A}$. We have the three polynomials $\mu_a(\lambda)$, $m_a(\lambda)$ and the characteristic polynomial $\chi_{\rho(a)}(\lambda)$ of the matrix $\rho(a)$. Since $m_a(a) = 0$, $\mu_a(\lambda) | m_a(\lambda)$. We also have $m_a(\lambda) | \chi_{\rho(a)}(\lambda)$ since $m_x(\lambda) | \chi_{\tilde{\rho}(x)}(\lambda)$. Also $\mu_a(\lambda)$ is the minimum polynomial of the matrix $\rho(a)$ so $\mu_a(\lambda)$ and $\chi_{\rho(a)}(\lambda)$ have the same irreducible factors. Hence $\mu_a(\lambda)$ and $m_a(\lambda)$ have the same irreducible factors and the same roots in $\bar{F}$.

We recall that $\mathscr{A}$ is *separable* if $\mathscr{A}_{\bar{F}} = M_{n_1}(\bar{F}) \oplus M_{n_2}(\bar{F}) \oplus \cdots \oplus M_{n_s}(\bar{F})$. It follows from the Hamilton–Cayley theorem that the degree of $M_n(F)$ is m and if $a \in M_m(F)$ then $m_a(\lambda) = \chi_a(\lambda)$. It follows easily that the degree of

$$\mathscr{A}_{\bar{F}} = M_{n_1}(\bar{F}) \oplus \cdots \oplus M_{n_s}(\bar{F})$$

and hence of $\mathscr{A}$ is $m = \sum_1^s n_i$ ([12], p. 228).

We shall call an element a of an algebra *separable* if $F[a]$ is a separable algebra. This is equivalent to: the minimum polynomial $\mu_a(\lambda)$ has distinct roots. We can now prove

1.11 Theorem. *Let $\mathscr{A}$ be separable of degree m over an infinite field F. Then the subset of $\mathscr{A}$ of elements a that are separable of degree m is non-vacuous Zariski open.*

Proof. Let x be a generic element of $\mathscr{A}$, $\delta(x)$ the discriminant of $m_x(\lambda)$. We claim that $\delta(x) \neq 0$. To see this we note that since $\mathscr{A}_{\bar{F}} = M_{n_1}(\bar{F}) \oplus \cdots \oplus M_{n_s}(\bar{F})$ there exist $m = \sum_1^s n_i$ non-zero orthogonal idempotents a_i in $\mathscr{A}_{\bar{F}}$ such that $\sum e_i = 1$. Let $\alpha_1, \ldots, \alpha_m$ be distinct in $\bar{F}$ and put $a = \sum \alpha_i e_i$. Then $\mu_a(\lambda) = \prod (\lambda - \alpha_i)$ is of degree m. Hence $m_a(\lambda) = \mu_a(\lambda)$ has distinct roots. Then $\delta(a) \neq 0$ and hence $\delta(x) \neq 0$. We have seen in Proposition 1.9 that the set of elements of $\mathscr{A}$ of degree m is non-vacuous open. The subset of these that are separable is defined by $\delta(a) \neq 0$. This is non-vacuous Zariski open. □

Theorem 1.11 gives a very natural proof of a classical result of Koethe's:

1.12 Theorem. *Any finite dimensional central simple algebra $\mathscr{A}$ has a finite dimensional separable splitting field.*

Proof. Since $\mathscr{A} = M_r(\mathscr{D})$ where $\mathscr{D}$ is a central division it suffices to prove the theorem for $\mathscr{A} = \mathscr{D}$ a division algebra. In this case if F is finite $\mathscr{D} = F$ by Wedderburn's theorem on the commutativity of finite division rings. Hence we may assume F infinite. In this case Theorem 1.11 shows that there exists a separable subfield K/F of $\mathscr{D}$ of degree $m = \deg \mathscr{D}$. Such a subfield is a splitting field (BA II, p. 224). □

2. Brauer Factor Sets

We recall that a finite dimensional associative algebra $\mathscr{A}/F$ is called a *Frobenius algebra* if there exists a non-degenerate bilinear form on $\mathscr{A}$ that is associative in the sense that

$$f(ab, c) = f(a, bc). \tag{2.1}$$

If $\mathscr{A} = \mathscr{A}_1 \oplus \mathscr{A}_2$ then $\mathscr{A}$ is Frobenius if and only if $\mathscr{A}_i$, $i = 1, 2$, is Frobenius. The tensor product of Frobenius algebras is Frobenius and any algebra $F[u]$ with a single generator is Frobenius (see Curtis and Reiner [7] or Jacobson [11]). If $\mathscr{A}$ is separable then $\mathscr{A}$ is Frobenius since the *reduced trace bilinear form* $t(a, b) = t(ab)$ is non-degenerate and associative. The following theorem has been proved by the author in [11]:

2.2 Theorem. *Let $\mathscr{A}$ be a central simple algebra of degree n, K a commutative Frobenius subalgebra of dimension n over F. Then there exists a $v \in \mathscr{A}$ such that $\mathscr{A} = KvK$.*

In particular this holds if K is a separable subalgebra with $[K:F] = n$. If F is infinite, by Theorem 1.11, there exists a $K = F[u]$ with u separable of degree n. The same can be seen readily also if F is finite (so $\mathscr{A} = M_n(F)$). We now assume that $K = F[u]$ is separable of degree n and we choose a v so that $\mathscr{A} = KvK$. Let $f(\lambda)$ be the minimum polynomial of u and E a splitting field over F of $f(\lambda)$ so $E = F(r_1, r_2, \ldots, r_n)$ where the r_i are distinct and $f(\lambda) = \prod(\lambda - r_i)$ in $E[\lambda]$. Consider the algebra $K_E = E[u] \cong E[\lambda]/(f(\lambda))$. In this algebra we have n non-zero orthogonal idempotents

$$e_i = \frac{(u - r_1)\cdots(u - r_{i-1})(u - r_{i+1})\cdots(u - r_n)}{(r_i - r_1)\cdots(r_i - r_{i-1})(r_i - r_{i+1})\cdots(r_i - r_n)} \tag{2.3}$$

such that $\sum e_i = 1$. Hence $K_E = \bigoplus_1^n Eu_i$. Now $\mathscr{A}_E$ is central simple of degree n over E and this contains K_E and hence the n orthogonal idempotents e_i. It follows that $\mathscr{A}_E = M_n(E)$ so E is a splitting field for $\mathscr{A}$. Thus we may regard $\mathscr{A}$ as an F-subalgebra of $M_n(E)$ such that $E\mathscr{A} = M_n(E)$. Also we may suppose that

$$u = \operatorname{diag}\{r_1, r_2, \ldots, r_n\}. \tag{2.4}$$

If $v = (v_{ij})$ then $u^k v u^l = (r_i^k r_j^l v_{ij})$ and since $\mathscr{A} = KvK$ the elements $u^k v u^l$, $0 \leq k$, $l \leq n - 1$, form a base for $\mathscr{A}/F$. Hence every element of $\mathscr{A}$ is a matrix

$$L = (l_{ij} v_{ij}), \tag{2.5}$$

where

$$l_{ij} = \sum_{k,l=1}^{n} a_{kl} r_i^{k-1} r_j^{l-1} \tag{2.6}$$

and the $a_{kl} \in F$ and are uniquely determined. Since $E\mathscr{A} = M_n(E)$ it is clear that every $v_{ij} \neq 0$.

Let $G = \operatorname{gal} E/F$. If $\sigma \in G$, $\sigma r_i = r_{i'}$, and σ is determined by the permutation

$i \rightsquigarrow i'$ of $\{1, 2, \ldots, n\}$. We denote this permutation by σ also, so we have $\sigma r_i = r_{\sigma i}$. If l_{ij}, $1 \le i, j \le n$, is defined by (2.6) then the l_{ij} satisfy

$$\sigma l_{ij} = l_{\sigma i, \sigma j}, \qquad \sigma \in G. \tag{2.7}$$

These conditions, which we shall call the *conjugacy conditions* on $l = (l_{ij})$, are also sufficient that the l_{ij} have the form (2.6); for we have

2.8 Lemma. *Let $l = (l_{ij})$ be a matrix of elements $l_{ij} \in E$ satisfying the conjugacy conditions (2.7). Then there exist $a_{kl} \in F$ such that (2.6) holds for all i, j.*

Proof. Let V be the Vandermonde matrix

$$V = \begin{bmatrix} 1 & r_1 & r_1^2 & \cdots & r_1^{n-1} \\ 1 & r_2 & r_2^2 & \cdots & r_2^{n-1} \\ \vdots & \vdots & \vdots & \cdots & \vdots \\ 1 & r_n & r_n^2 & \cdots & r_n^{n-1} \end{bmatrix}. \tag{2.9}$$

Then V is invertible. Hence there exists a unique matrix $a = (a_{ij}) \in M_n(E)$ such that

$$Va({}^tV) = l. \tag{2.10}$$

This matrix relation is equivalent to the equations (2.6). Applying $\sigma \in G$ to these equations we obtain

$$l_{\sigma i, \sigma j} = \sum_{k,l} (\sigma a_{kl}) r_{\sigma i}^{k-1} r_{\sigma j}^{l-1},$$

or $l_{ij} = \sum_{k,l} (\sigma a_{kl}) r_i^{k-1} r_j^{l-1}$. By the uniqueness of a we have $\sigma a_{kl} = a_{kl}$ for every $\sigma \in G$. Hence $a_{kl} \in F$. $\square$

(The foregoing proof is due to Walter Feit.)

We now put $L = (l_{ij} v_{ij})$, $L' = (l'_{ij} v_{ij})$ where the l_{ij} and l'_{ij} satisfy the conjugacy conditions, so $L, L' \in \mathscr{A}$. Then $LL' = L'' = (l''_{ij} v_{ij})$ where

$$l''_{ij} = \sum l_{ik} c_{ikj} l'_{kj}, \tag{2.11}$$

$$c_{ikj} = v_{ik} v_{kj} v_{ij}^{-1}. \tag{2.12}$$

2.13 Lemma. *The c_{ijk} satisfy*

$$\sigma c_{ijk} = c_{\sigma i, \sigma j, \sigma k}, \tag{2.14}$$

$$c_{ijk} c_{ikl} = c_{ijl} c_{jkl}. \tag{2.15}$$

Proof. Apply σ to (2.11) to obtain $l''_{\sigma i, \sigma j} = \sum_k l_{\sigma i, \sigma k} (\sigma c_{ikj}) l'_{\sigma k, \sigma j}$. On the other hand, $l''_{\sigma i, \sigma j} = \sum_k l_{\sigma i, \sigma k} c_{\sigma i, \sigma j, \sigma k} l'_{\sigma k, \sigma j}$. Hence we have

$$\sum_k l_{\sigma i, \sigma k} (\sigma c_{ikj} - c_{\sigma i, \sigma k, \sigma j}) l'_{\sigma k, \sigma j} = 0,$$

or

$$\sum_k l_{ik} d_{ikj} l'_{ij} = 0, \qquad d_{ikj} = \sigma c_{\sigma^{-1} i, \sigma^{-1} k, \sigma^{-1} j} - c_{ikj} \tag{2.16}$$

and these relations hold for all l_{ik}, l'_{jk} satisfying the conjugacy conditions. We can write also

(2.17) $$\sum_k l_{ik} e_{ikj}(v_{kj} l'_{kj}) = 0, \qquad e_{ikj} = d_{ikj} v_{kj}^{-1}.$$

Now $M_n(E) = E\mathscr{A}$. Hence taking a suitable E-linear combination of the matrices in $\mathscr{A}$ we obtain a matrix whose jth column is $(0, \ldots, 0, 1, 0, \ldots, 0)$ where the 1 is in any chosen position. Using this linear combination of the relations (2.17) we obtain $l_{ik} e_{ikj} = 0$ for all k. Then $e_{ikj} = 0$ and $d_{ikj} = 0$ which is (2.14). Now (2.15) follows by direct verification using the definition (2.12). □

We now define a *Brauer factor set* c to be an indexed set of elements $c_{ijk} \in E^*$, $1 \le i, j, k \le n$, such that

(i) $$\sigma c_{ijk} = c_{\sigma i, \sigma j, \sigma k},$$

(ii) $$c_{ijk} c_{ikl} = c_{ijl} c_{jkl}.$$

The foregoing lemma states that the c_{ijk} defined by (2.12) from the element $v = (v_{ij})$ constitute a Brauer factor set. We shall call (i) the conjugacy conditions on the c_{ijk}. We note that these imply that

(iii) $$c_{ijk} \in F(r_i, r_j, r_k).$$

For, if $\sigma \in \operatorname{Gal} E/F(r_i, r_j, r_k)$ then $\sigma i = i, \sigma j = j, \sigma k = k$ and hence, by (i), $\sigma c_{ijk} = c_{ijk}$. Since this holds for every $\sigma \in \operatorname{Gal} E/F(r_i, r_j, r_k)$, $c_{ijk} \in F(r_i, r_j, r_k)$ by the Galois correspondence. If we put $i = j = k$ and $j = k = l$ successively in (ii) we obtain

(iv) $$c_{iij} = c_{iii} = c_{jii}.$$

We have seen that if we define $c_{ijk} = v_{ij} v_{jk} v_{ik}^{-1}$ then $c = \{c_{ijk}\}$ is a Brauer factor set. Here $v = (v_{ij})$ was any element of $\mathscr{A}$ such that $\mathscr{A} = KvK$. We now observe that c is independent of the imbedding of $\mathscr{A}$ in $M_n(E)$ provided that

$$u = \operatorname{diag}\{r_1, r_2, \ldots, r_n\}$$

in the imbedding. For, if we have a second imbedding with this property then it follows from the Skolem–Noether theorem and the fact that the only matrices that commute with u are diagonal matrices that in the second imbedding we have $v = (d_i v_{ij} d_j^{-1})$ where $d_i \in E^*$. Then

$$(d_i v_{ik} d_k^{-1})(d_k v_{kj} d_j^{-1})(d_i v_{ij} d_j^{-1})^{-1} = v_{ik} v_{kj} v_{ij}^{-1} = c_{ijk}.$$

We shall now normalize v so that the corresponding factor set c is *reduced* in the sense that every $c_{iii} = 1$. By (iv) this implies $c_{iij} = 1 = c_{jii}$ for all i, j. We remark that if $f(\lambda)$ is irreducible or, equivalently, K is a field then c is reduced if $c_{111} = 1$. For, in this case the permutation group of the r_i determined by G is transitive. Then $c_{111} = 1$ implies $c_{iii} = 1$. We now note that, by (2.12), $c_{iii} = v_{ii}$ so $\sigma v_{ii} = v_{\sigma i, \sigma i}$, $\sigma \in G$. Hence if we put $l_{ii} = v_{ii}^{-2}$, $l_{ij} = 0$ if $i \ne j$ then the conjugacy conditions hold for the l_{ij} so

(2.18) $$y = \operatorname{diag}\{v_{11}^{-1}, v_{22}^{-1}, \ldots, v_{nn}^{-1}\} \in \mathscr{A}.$$

Since y commutes with u, $y \in F[u]$ and we can replace v by yv. This normalization permits us to assume $v_{ii} = 1$ and hence $c_{iii} = 1$, that is, c is reduced.

We can now prove

2.19 Theorem. *Let $K = F[u]$ be finite dimensional separable, $f(\lambda)$ the minimum polynomial of u over F and let $E = F(r_1, \ldots, r_n)$ be the splitting field of $f(\lambda)$ over F where $f(\lambda) = \prod (\lambda - r_i)$ in $E[\lambda]$. Suppose $c = \{c_{ijk}\}$ is a reduced Brauer factor set with values in E^* and let $\mathscr{B}(K, c)$ denote the subset of $M_n(E)$ of matrices $l = (l_{ij})$ such that $\sigma l_{ij} = l_{\sigma i, \sigma j}$, $\sigma \in G = \mathrm{Gal}\, E/F$. Then $\mathscr{B}(K, c)$ is an F-subspace of $M_n(E)$ and if we define a product $l_c l'$ for $l = (l_{ij})$, $l' = (l'_{ij}) \in \mathscr{B}(K, c)$ as $l'' = (l''_{ij})$ where*

$$l''_{ij} = \sum_k l_{ik} c_{ikj} l'_{kj}, \tag{2.20}$$

then $\mathscr{B}(K, c)$ becomes a central simple associative algebra of degree n over F containing a subalgebra isomorphic to K. Moreover, the map

$$l = (l_{ij}) \rightsquigarrow L = (c_{ij1} l_{ij}) \tag{2.21}$$

is an isomorphism of $\mathscr{B}(K, c)$ with an F-subalgebra $\mathscr{A}$ of $M_n(E)$.

Conversely, every central simple algebra of degree n containing K as subalgebra can be obtained by this construction.

Proof. It is clear that $\mathscr{B} = \mathscr{B}(K, c)$ is an F-subspace of $M_n(E)$ and if l''_{ij} is defined by (2.20) then

$$\begin{aligned} \sigma l''_{ij} &= \sum_k (\sigma l_{ik})(\sigma c_{ikj})(\sigma l'_{kj}) \\ &= \sum_k l_{\sigma i, \sigma k} c_{\sigma i, \sigma k, \sigma j} l'_{\sigma k, \sigma j} \\ &= \sum_k l_{\sigma i, k} c_{\sigma i, k, \sigma j} l'_{k, \sigma j} \\ &= l''_{\sigma i, \sigma j}. \end{aligned}$$

Hence $\mathscr{B}$ is closed under the "c-multiplication." Consider the map defined by (2.21). Evidently this is F-linear and injective. The (i, j)-entry of the matrix product $(c_{ij1} l_{ij})(c_{ij1} l'_{ij})$ is

$$\begin{aligned} \sum_k c_{ik1} c_{kj1} l_{ik} l'_{kj} &= \sum_k c_{ij1} c_{ikj} l_{ik} l'_{kj} \qquad \text{(by (ii))} \\ &= c_{ij1} l''_{ij}. \end{aligned}$$

Hence the map is a homomorphism for multiplication. The image $\mathscr{A} = \{L\}$ of $\mathscr{B}$ is an F-subspace of $M_n(E)$ closed under multiplication. Observe next that since $c_{ii1} = 1$, any diagonal matrix satisfying the conjugacy conditions is fixed under (2.21). Then $1 \in \mathscr{A}$ and is the unit of $\mathscr{A}$ and of $\mathscr{B}$. Hence $\mathscr{A}$ is an F-subalgebra of $M_n(E)$.

We note next that $r = \mathrm{diag}\{r_1, r_2, \ldots, r_n\} \in \mathscr{A}$ and $F[r]$ is a subalgebra of $\mathscr{A}$ isomorphic to K. Next let $l_{ij} = 1$ for all i, j and let s be the corresponding matrix $(c_{ij1} l_{ij}) = (c_{ij1})$. Note that every entry of s is $\neq 0$. Now every matrix unit $e_{ii} \in E[r]$, and since $e_{ii}se_{jj}$ is a non-zero multiple of e_{ij} it is clear that $E\mathscr{A} = M_n(E)$. Since $M_n(E)$ is simple $\mathscr{A}$ contains no nilpotent ideals $\neq 0$ and $\mathscr{A}$ is not a direct sum of more than one non-zero ideal. Hence by the Wedderburn structure theory, $\mathscr{A}$ is simple. Any element of the center of $\mathscr{A}$ is in the center of $M_n(E)$ and so is a scalar matrix. Such an element has pre-image under (2.21) that is a diagonal matrix $\text{diag}\{l_1, \ldots, l_n\}$. The conditions $\sigma l_i = l_{\sigma i}$ and $\text{diag}\{l_1, \ldots, l_n\}$ is a scalar matrix imply that this element is in $F1$. Hence $\mathscr{A}$ is central simple. Then $M_n(E) \cong E \otimes_F \mathscr{A}$ (BA II, Theorem 4.7) and consequently $\mathscr{A}$ has degree n. The isomorphism of $\mathscr{B}(K, c)$ with $\mathscr{A}$ now implies that $\mathscr{B}(K, c)$ is central simple of degree n and $\mathscr{B}(K, c)$ contains a subalgebra isomorphic to K.

Conversely, assume $\mathscr{A}$ is central simple of degree n containing $K = F[u]$. We have seen that $\mathscr{A} = KvK$ and we can identify $\mathscr{A}$ with the F-subalgebra of matrices $(v_{ij}l_{ij})$ where $v = (v_{ij})$ has all its entries $\neq 0$ and (l_{ij}) satisfies the conjugacy conditions. Moreover, if we define $c_{ikj} = v_{ik}v_{kj}v_{ij}^{-1}$ then $c = \{c_{ijk}\}$ is a Brauer factor set. By normalizing v we may assume c is reduced. Now we have $(v_{ij}l_{ij})(v_{ij}l'_{ij}) = (v_{ij}l''_{ij})$ where l''_{ij} is given by (2.20). Hence the map $(l_{ij}) \rightsquigarrow (v_{ij}l_{ij})$ is an isomorphism of $(\mathscr{B}, c)$ onto $\mathscr{A}$. □

We shall now determine the elements $w \in \mathscr{A}$ such that $\mathscr{A} = KwK$. We claim that these are just the elements $w = (l_{ij}v_{ij})$ such that every $l_{ij} \neq 0$. We have seen that $E[u] = D = \sum Ee_{ii}$. It is clear that $DwD = M_n(E)$ for a matrix $w = (w_{ij})$ if and only if every $w_{ij} \neq 0$. On the other hand, $D = EK$ and hence $DwD = EKwEK = EKwK$. Now if $w \in \mathscr{A}$ then $KwK \subset \mathscr{A}$ and hence $EKwK = E \otimes_F KwK$. Hence $\mathscr{A} = KwK$ for $w \in \mathscr{A}$ if and only if $w = (l_{ij}v_{ij})$ with every $l_{ij} \neq 0$.

We have associated with an element $v \in \mathscr{A}$ such that $\mathscr{A} = KwK$ a factor set $c = \{c_{ijk}\}$ where $c_{ijk} = v_{ij}v_{jk}v_{ik}^{-1}$for $v = (v_{ij})$. If $w = (l_{ij}v_{ij})$ where the l_{ij} satisfy the conjugacy conditions and every $l_{ij} \neq 0$ then the Brauer factor set determined by w is $c' = \{c'_{ijk}\}$ where

$$c'_{ijk} = l_{ij}l_{jk}l_{ik}^{-1}c_{ijk}. \tag{2.22}$$

Two Brauer factor sets related in this way by l_{ij} satisfying the conjugacy conditions are called *associates*. These constitute an equivalence class. We denote the equivalence class of the Brauer factor sets all of which $c_{ijk} = 1$ by 1 and the relation of associativeness by $\sim$.

3. Condition for Split Algebra. The Tensor Product Theorem

We retain the notations of the last section. We prove first

3.1 Theorem. *$\mathscr{B}(K, c) \sim 1$ in the Brauer group* Br(F) *(that is, $\mathscr{B}(K, c) \cong M_n(F)$) if and only if $c \sim 1$.*

Proof. Suppose $c \sim 1$. Then we may assume every $c_{ijk} = 1$. Hence the subalgebra $\mathscr{A}$ of $M_n(E)$ isomorphic to $\mathscr{B}(K, c)$ contains the matrix v all of whose entries are 1. This matrix has rank 1 and hence the left ideal $M_n(E)v$ of $M_n(E)$ is minimal and so is n dimensional over E. It follows that $[\mathscr{A}v:F] = n$. Then $\mathscr{A}$ has a representation by $n \times n$ matrices over F determined by the module $\mathscr{A}v$. It follows that $\mathscr{A} \cong M_n(F)$. Conversely, suppose $\mathscr{B}(K, c) \cong M_n(E)$. Then $\mathscr{B}(K, c) \cong \mathscr{B}(K, 1)$. We have shown in [11] that if $\mathscr{A}$ is central simple of degree n and K_1 and K_2 are isomorphic commutative Frobenius subalgebras of $\mathscr{A}$ with $[K_i:F] = n$ then any isomorphism of K_1 onto K_2 can be extended to an inner automorphism of $\mathscr{A}$. Hence if $\mathscr{B}(K, c) \cong \mathscr{B}(K, 1)$ then we may assume that we have an isomorphism between these algebras that is the identity map on K. Let $\mathscr{A}_1$ and $\mathscr{A}_2$ be the subalgebras of $M_n(E)$ isomorphic to $\mathscr{B}(K, 1)$ and $\mathscr{B}(K, c)$, respectively, so that $\mathscr{A}_i$ contains the matrix $u = \text{diag}\{r_1, r_2, \ldots, r_n\}$, $\mathscr{A}_1$ contains the matrix v_1 all of whose entries are 1 and $\mathscr{A}_2$ contains $v_2 = (v_{ij})$ so that $c_{ijk} = v_{ik}v_{kj}v_{ij}^{-1}$. Then $\mathscr{A}_1 = F[u]v_1F[u]$, $\mathscr{A}_2 = F[u]v_2F[u]$ and we have an isomorphism η of $\mathscr{A}_2$ onto $\mathscr{A}_1$ that is the identity on u. Then $w_1 = \eta(v_2)$ satisfies $\mathscr{A}_1 = F[u]w_1F[u]$ and we have seen that the Brauer factor set determined by w_1 is c. Since that determined by v_1 is 1 we have $c \sim 1$. □

We consider next the tensor product of two central simple algebras $\mathscr{A}_i$, $i = 1, 2$, of degree n containing $K = F[u]$ as subalgebra. Let v_i be an element of $\mathscr{A}_i$ such that $\mathscr{A}_i = Kv_iK$ and let $v_i = (v_{jk}^{(i)})$ in an imbedding of $\mathscr{A}_i$ in $M_n(E)$. The algebra $\mathscr{A}_1 \otimes_F \mathscr{A}_2$ contains $K \otimes_F K$. We have the exact sequence of algebra homomorphisms

$$K \otimes_F K \xrightarrow{\nu} K \to 0, \tag{3.2}$$

where $\nu: \sum a_i \otimes b_i \rightsquigarrow \sum a_i b_i$. Since $K \otimes_F K$ is semi-simple we have

$$K \otimes_F K = (K \otimes_F K)e \oplus (K \otimes_F K)(1 - e), \tag{3.3}$$

where e is an idempotent and $(K \otimes_F K)(1 - e) = \ker \nu$. Then $(K \otimes K)e \cong (K \otimes K)/\ker \nu \cong K$. Moreover, since $a \otimes 1 - 1 \otimes a \in \ker \nu$ for $a \in K$, we have

$$(a \otimes 1)e = (1 \otimes a)e, \qquad a \in K. \tag{3.4}$$

We now consider the algebra

$$\mathscr{A} = e(\mathscr{A}_1 \otimes_F \mathscr{A}_2)e. \tag{3.5}$$

Since $\mathscr{A}_1$ and $\mathscr{A}_2$ are central simple so is $\mathscr{A}_1 \otimes_F \mathscr{A}_2$ and hence so is $\mathscr{A}$. Moreover, $\mathscr{A}$ and $\mathscr{A}_1 \otimes_F \mathscr{A}_2$ determine the same element of the Brauer group Br(F) and $\mathscr{A}$ contains $e(K \otimes_F K)e \cong K$ which we can identify with K. Then we have

3.6 Theorem. *$\mathscr{A}$ is of degree n containing K, and a Brauer factor set associated with $\mathscr{A}$ is $c^{(1)}c^{(2)} = \{c_{jkl}^{(1)}c_{jkl}^{(2)}\}$ where $c^{(i)} = \{c_{jkl}^{(i)}\}$ is a Brauer factor set associated with $\mathscr{A}_i$.*

Proof. If $a^{(1)} = (\alpha_{ij}^{(1)})$, $a^{(2)} = (\alpha_{ij}^{(2)}) \in M_n(E)$ we define

$$a^{(1)} \otimes a^{(2)} = \begin{bmatrix} \alpha_{11}^{(1)}a^{(2)} & \alpha_{12}^{(1)}a^{(2)} & \cdots & \alpha_{1n}^{(1)}a^{(2)} \\ \alpha_{21}^{(1)}a^{(2)} & \alpha_{22}^{(1)}a^{(2)} & \cdots & \alpha_{2n}^{(1)}a^{(n)} \\ \vdots & \vdots & \cdots & \vdots \\ \alpha_{n1}^{(1)}a^{(2)} & \alpha_{n2}^{(1)}a^{(2)} & \cdots & \alpha_{nn}^{(1)}a^{(n)} \end{bmatrix} \tag{3.7}$$

and we use this tensor multiplication of matrices to obtain an imbedding of $\mathscr{A}_1 \otimes_F \mathscr{A}_2$ in $M_{n^2}(F)$. Since $u = \text{diag}\{r_1, r_2, \ldots, r_n\}$ in $M_n(E)$ it is clear that the matrix for any $a \in K \otimes_F K$ in $M_{n^2}(E)$ is a diagonal matrix. Hence the matrix for e is diagonal with entries 0 and 1. Also we have

$$(3.8) \qquad u \otimes 1 = \begin{bmatrix} r_1 1_n & & & 0 \\ & r_2 1_n & & \\ 0 & & \ddots & \\ & & & r_n 1_n \end{bmatrix},$$

$$(3.9) \qquad 1 \otimes u = \begin{bmatrix} u & & & 0 \\ & u & & \\ 0 & & \ddots & \\ & & & u \end{bmatrix},$$

Hence the condition (3.4) for $a = u$ implies that all the diagonal entries of e are 0 with the exception of those in the positions $1, n + 2, 2n + 3, \ldots, n^2$. This implies that $eM_{n^2}(E)e$ has degree $\leq n$ and hence the degree of $\mathscr{A} = e(\mathscr{A}_1 \otimes \mathscr{A}_2)e$ is $\leq n$. On the other hand, this degree is $\geq n$ since $\mathscr{A} \supset K$. Hence $\mathscr{A}$ has degree n and the diagonal entries of e in the positions $1, n + 2, \ldots, n^2$ are 1 and the remaining ones are 0. Then matrix $e(v_1 \otimes v_2)e$ in $M_{n^2}(E)$ has non-zero entries only in the

$$((k - 1)n + k, (l - 1)n + l)$$

positions $1 \leq k, l \leq n$, and the entry in this position is $v_{kl}^{(1)}v_{kl}^{(2)}$.

By performing a similarity transformation by a permutation matrix and cutting down to a diagonal block we obtain an imbedding of $\mathscr{A}$ in $M_n(E)$ in which $u = \operatorname{diag}\{r_1, \ldots, r_n\}$ and $v = e(v_1 \otimes v_2)e = (v_{kl}^{(1)}v_{kl}^{(2)})$. Since all the entries of v are $\neq 0$, $M_n(E) = (\sum Ee_{ii})v(\sum Ee_{ii})$ and hence $\mathscr{A} = KvK$. It follows that we can use v to determine a Brauer factor set for $\mathscr{A}$. Evidently this set is $c^{(1)}c^{(2)}$. □

4. The Brauer Group Br(K/F)

From now on we assume the base field F is infinite. As before, let K be a finite dimensional commutative separable algebra over F. If $\bar{F}$ is the algebraic closure of F then $K_{\bar{F}} = \bar{F}e_1 \oplus \cdots \oplus \bar{F}e_n$ where the e_i are orthogonal idempotents and $n = [K:F]$. It follows that the degree of $K = \deg K_{\bar{F}} = n$. Hence, by Theorem 1.11, $K = F[u]$ where u is separable with minimum polynomial $f(\lambda)$ of degree n. We shall say that an extension field E/F *splits* K if $K_E = Ee_1 \oplus \cdots \oplus Ee_n$ where the e_i are orthogonal idempotents and we call E a *splitting field* for K if E splits K and no proper subfield of E splits K. It is readily seen that E is a splitting field for K/F if and only if E is a splitting field in the usual sense for the polynomial $f(\lambda)$. Hence any two splitting fields E/F and E'/F of K/F are isomorphic.

Now let E/F be a splitting field for K/F where $K = F[u]$ and $f(\lambda)$ is the minimum polynomial of u. Then E is a splitting field of $f(\lambda)$. For each root r of $f(\lambda)$ we have a homomorphism α of K/F into E/F such that $u \rightsquigarrow r$. In this way we obtain $n = [K:F]$ homomorphisms of K/F into E/F such that $\alpha_i u = r_i$ where $f(\lambda) = \prod (\lambda - r_i)$ in $<[\lambda]$. Moreover, this gives all the homomorphisms of K/F into E/F.

Thus

(4.1) $$M = \{\alpha_1, \alpha_2, \ldots, \alpha_n\}$$

is the set of homomorphisms of K/F into E/F. If $\sigma \in G = \text{Gal } E/F$ then $\sigma\alpha_i \in M$. In fact, we have $\sigma\alpha_i u = \sigma r_i = r_{\sigma i}$ so $\sigma\alpha_i = \alpha_{\sigma i}$.

Now let $c = \{c_{ijk}\}$ be a Brauer factor set. We can regard this as a map c of $M \times M \times M$into E^* such that

(4.2) $$c: (\alpha_i, \alpha_j, \alpha_k) \rightsquigarrow c_{ijk}.$$

Accordingly, we write $c(\alpha_i, \alpha_j, \alpha_k)$ for c_{ijk}. Then the defining conditions on the c_{ijk} are first that

(4.3) $$\sigma c(\alpha_i, \alpha_j, \alpha_k) = \sigma c_{ijk} = c_{\sigma i, \sigma j, \sigma k} = c(\sigma\alpha_i, \sigma\alpha_j, \sigma\alpha_k)$$

or, independently of the indexing,

(4.3′) $$\sigma c(\alpha, \beta, \gamma) = c(\sigma\alpha, \sigma\beta, \sigma\gamma), \qquad \alpha, \beta, \gamma \in M.$$

We shall now call these conditions *homogeneity* and, more generally, if $g: \overbrace{M \times \cdots \times M}^{r} \to E$ or E^* then E is *homogeneous* if

(4.4) $$g(\sigma\alpha, \sigma\beta, \ldots, \sigma\varepsilon) = \sigma g(\alpha, \beta, \ldots, \varepsilon)$$

for $\alpha, \beta, \ldots, \varepsilon \in M$. In addition to this condition on c we have

(4.5) $$c(\alpha, \beta, \gamma)c(\alpha, \gamma, \delta) = c(\alpha, \beta, \delta)c(\beta, \gamma, \delta)$$

for α, β, γ, $\delta \in M$. c is reduced if $c(\alpha, \alpha, \alpha) = 1$ for all $\alpha \in M$. This implies that $c(\beta, \alpha, \alpha) = 1 = c(\alpha, \alpha, \beta)$ for all α, β. If K is a field then c is reduced if $c(\alpha, \alpha, \alpha) = 1$ for a single $\alpha \in M$.

Similarly, a matrix $l = (l_{ij}) \in M_n(E)$ can be regarded as a map $(\alpha_i, \alpha_j) \rightsquigarrow l_{ij}$. The usual matrix product of l and l' can then be defined by $ll'(\alpha, \beta) = \sum_{\gamma \in M} l(\alpha, \gamma)l'(\gamma, \beta)$. Homogeneity of l as map of $M \times M \to E$ is equivalent to the conjugacy conditions $l_{\sigma i, \sigma j} = \sigma l_{ij}$.

We can now re-state Theorem 2.19 in the following way:

4.6 Theorem. *Let K/F be a finite dimensional separable commutative algebra, E/F a splitting field for K/F, c a reduced Brauer factor set with values in E^*. Let $\mathscr{B}(K, c)$ denote the F-space of homogeneous maps of $M \times M$ into E and define a product in $\mathscr{B}(K, c)$ by*

(4.7) $$ll'(\alpha, \beta) = \sum_{\gamma \in M} l(\alpha, \gamma)c(\alpha, \gamma, \beta)l'(\gamma, \beta)$$

for $l, l' \in \mathscr{B}(K, c)$. Then $\mathscr{B}(K, c)$ becomes a central simple algebra of degree $n = [K:F]$ containing a subalgebra isomorphic to K. Moreover, for any fixed $\gamma \in M$ the map $l \rightsquigarrow L$ where

(4.8) $$L(\alpha, \beta) = c(\alpha, \beta, \gamma)l(\alpha, \beta)$$

is an isomorphism of $\mathscr{B}(K, c)$ with an F-subalgebra $\mathscr{A}$ of the matrix algebra of maps of $M \times M$ into E.

Conversely, any central simple algebra of degree n containing K as a subalgebra can be obtained in this way.

We shall call $\mathscr{B}(K, c)$ the *Brauer algebra determined by the Brauer factor set c*. The condition that K is a commutative separable subalgebra of dimension equal to the degree is equivalent to two other conditions given in

4.9 Theorem. *Let $\mathscr{A}$ be central simple of degree n over F, K/F a commutative separable subalgebra of $\mathscr{A}$. Then the following conditions on K are equivalent:*

(i) $[K:F] = n$;
(ii) *K is a maximal commutative separable subalgebra of* $\mathscr{A}$;
(iii) $\mathscr{A}^K = K$ *for the centralizer* $\mathscr{A}^K$ *of K in* $\mathscr{A}$.

Proof. (i) $\Rightarrow$ (ii). Suppose L is a commutative separable subalgebra of $\mathscr{A}$ containing K. Then $L = F[v]$ and the degree of the minimum polynomial of $v \leq \deg \mathscr{A} = [K:F]$. Hence $[L:F] \leq [K:F]$ so $L = K$.

(ii) $\Rightarrow$ (iii). Let K be a maximal commutative subalgebra of $\mathscr{A}$. Then $K \subset \mathscr{A}^K$. Now $\mathscr{A}^K$ is separable. For, if $\bar{F}$ is the algebraic closure of F then $\mathscr{A}_{\bar{F}} = M_n(\bar{F})$, $K_{\bar{F}} = \bar{F}e_i \oplus \cdots \oplus \bar{F}e_m$ where the e_i are non-zero orthogonal idempotents such that $\sum e_i = 1$. Then $(\mathscr{A}^K)_{\bar{F}} \cong \mathscr{A}_{\bar{F}}^{K_{\bar{F}}} = M_n(\bar{F})^{\sum \bar{F}e_i}$. It is clear that the last algebra is a direct sum of algebras $M_{n_i}(\bar{F})$. Hence $\mathscr{A}^K$ is separable. Then the center of $\mathscr{A}^K$ is separable and since it contains K it coincides with K by the maximality of K. Now $\mathscr{A}^K = \mathscr{A}_1 \oplus \cdots \oplus \mathscr{A}_s$ where $\mathscr{A}_i$ is separable with separable center K_i and $K = K_1 + \cdots + K_s$. Suppose for some i, $\mathscr{A}_i \supsetneqq K_i$. If $\mathscr{A}_i$ is not a division algebra then $\mathscr{A}_i$ contains $m \geq 2$ non-zero orthogonal idempotents f_j such that $\sum f_j = 1_i$ the unit of $\mathscr{A}_i$. Then $K_1 + \cdots + K_{i-1} + \sum K_i f_j + K_{i+1} + \cdots + K_s$ is a commutative separable subalgebra of $\mathscr{A}$ properly containing K contrary to the maximality of K. The same conclusion holds if $\mathscr{A}_i$ is a division algebra since in this case $\mathscr{A}_i$ contains a separable subfield properly containing K_i. These contradictions show that $\mathscr{A}_i = K_i$ for every i and hence $\mathscr{A}^K = K$.

(iii) $\Rightarrow$ (i) Suppose $\mathscr{A}^K = K$. Then $\mathscr{A}_{\bar{F}}^{K_{\bar{F}}} = K_{\bar{F}}$ for $\bar{F}$ the algebraic closure of F and hence $M_n(\bar{F})^{\sum_1^m \bar{F}e_i} = \sum \bar{F}e_i$ where the e_i are non-zero orthogonal idempotents such that $\sum e_i = 1$ and $m = [K:F]$. It follows that $m = n$ and $[K:F] = n$. □

The Brauer factor sets (regarded as maps of $M \times M \times M$ into E^*) form a group under multiplication of images in E^*. This contains the subgroup of factor sets such that

$$c(\alpha, \beta, \gamma) = l(\alpha, \beta)l(\beta, \gamma)l(\alpha, \gamma)^{-1}, \tag{4.10}$$

where $l: M \times M \to E^*$ is homogeneous. We can form the factor group which we shall denote as $H^2(K/F)$. If c is a factor set then l defined by $l(\alpha, \alpha) = c(\alpha, \alpha, \alpha)^{-1}$, $l(\alpha, \beta) = 1$ if $\alpha \neq \beta$ is homogeneous and $c(\alpha, \beta, \gamma)l(\alpha, \beta)l(\beta, \gamma)l(\alpha, \gamma)^{-1}$ is reduced. It follows that $H^2(K, F)$ is the factor group of the group of reduced Brauer factor sets with respect to its subgroup of reduced Brauer factor sets of the form (4.10).

We now let $\mathrm{Br}(K/F)$ denote the subset of the Brauer group $\mathrm{Br}(F)$ consisting of the classes $\{\mathscr{B}(K, c)\}$. If K is a field then we have the homomorphism of Brauer groups defined by $\{A\} \rightsquigarrow \{A_K\}$. The kernel is the subgroup of classes $\{A\}$ having K as splitting field. By the Brauer–Noether theorem on splitting fields, a central simple algebra has K as splitting field if and only if it is similar to an $\mathscr{A}$ containing K such as subfield such that $\mathscr{A}^K = K$ (BA II, p. 221). Thus it is clear that if K is a field then $\mathrm{Br}(K/F)$ is the kernel of the homomorphism of $\mathrm{Br}(F)$ into $\mathrm{Br}(K)$. We shall now prove for arbitrary commutative separable K the following;

4.11 Theorem. $\mathrm{Br}(K/F)$ *is a subgroup of* $\mathrm{Br}(F)$ *isomorphic to* $H^2(K/F)$.

Proof. We have the surjective map $c \rightsquigarrow \{\mathscr{B}(K, c)\}$ of the group of reduced Brauer factor sets with values in E^* onto $\mathrm{Br}(K/F)$. Now let $c^{(1)}$ and $c^{(2)}$ be reduced Brauer factor sets. Then it follows from Theorem 3.6 that $\mathscr{B}(K, c^{(1)}) \otimes_F \mathscr{B}(K, c^{(2)}) \sim \mathscr{B}(K, c^{(1)}c^{(2)})$. This implies that $\mathrm{Br}(K/F)$ is a subgroup of $\mathrm{Br}(F)$ and $c \rightsquigarrow \{\mathscr{B}(K, c)\}$ is a homomorphism of the group of reduced Brauer factor sets onto $\mathrm{Br}(K/F)$. By Theorem 3.1 the kernel of this homomorphism is the group of reduced $c \sim 1$. Hence $\mathrm{Br}(K/F) \cong H^2(K/F)$. □

We also have the following generalization of the theorem of Speiser–Noether that $H^1(G, E^*) = 1$ for G the Galois group of E/F.

4.12 Theorem. *Let K be a finite dimensional commutative separable algebra over F, E/F a splitting field for K/F, $M = \{\alpha\}$ the set of homomorphisms of K/F into E/F. Let $(\alpha, \beta) \rightsquigarrow b(\alpha, \beta)$ be a homogeneous map of $M \times M$ into E^* such that*

$$b(\alpha, \beta)b(\beta, \gamma) = b(\alpha, \gamma) \tag{4.13}$$

for $\alpha, \beta, \gamma \in M$. Then there exists an invertible $a \in K$ such that

$$b(\alpha, \beta) = (\alpha a)(\beta a)^{-1}. \tag{4.14}$$

Proof. Consider the algebra $\mathscr{B}(K, 1)$ which is the F-space of homogeneous maps of $M \times M$ into E with multiplication defined by $ll'(\alpha, \beta) = \sum_{\gamma \in M} l(\alpha, \gamma)l'(\gamma, \beta)$. For $a \in K$ we define a homogeneous map a' of $M \times M$ into E by $a'(\alpha, \alpha) = \alpha a$, $a'(\alpha, \beta) = 0$ if $\alpha \neq \beta$. Then $a \rightsquigarrow a'$ is a homomorphism of K into $\mathscr{B}(K, 1)$. This is a monomorphism since $K \otimes_F E = Ee_1 \oplus \cdots \oplus Ee_n$ where the e_i are orthogonal idempotents and for any $a \in K$, $a = \sum (\alpha_i a)e_i$ where $\alpha_i a \in E$. Then $\alpha_i \in M$ and if $\alpha_i a = 0$ for all i, $a = 0$. Thus we can identify K with its image in $\mathscr{B}(K, 1)$ and write a for a'. Then $\mathscr{B}(K, 1)^K = K$ by Theorem 4.9. We now consider the map $\eta: l \rightsquigarrow l'$ where $l'(\alpha, \beta) = l(\alpha, \beta)b(\alpha, \beta)$ for $l \in \mathscr{B}(K, c)$. The condition (4.13) implies that η is an automorphism of $\mathscr{B}(K, 1)$. Moreover, (4.13) gives $b(\alpha, \alpha)^2 = b(\alpha, \alpha)$ so $b(\alpha, \alpha) = 1$. Hence $\eta a = a$ for $a \in K$. It follows from the Skolem–Noether theorem and $\mathscr{B}(K, 1)^K = K$ that there exists an invertible $a \in K$ such that $\eta = I_a$ the inner automorphism $x \rightsquigarrow axa^{-1}$. Now let v be defined by $v(\alpha, \beta) = 0$ for all α, β. Then v is homogeneous and $v' = \eta v$ satisfies $v'(\alpha, \beta) = b(\alpha, \beta)$. Since $(ava^{-1})(\alpha, \beta) = (\alpha a)(\beta a)^{-1}$ we have $b(\alpha, \beta) = (\alpha a)(\beta a)^{-1}$ for $\alpha, \beta \in M$. □

5. Crossed Products

We shall now specialize to the case $E = K$, $M = G = \text{Gal}\, E/F$ in the foregoing considerations. In this case one has the crossed product representation, due to Emmy Noether, of an algebra $\mathscr{A}$ containing E and having degree $n = [E:F]$. Let $\sigma \in G$ then σ can be extended to an inner automorphism I_{u_σ} of $\mathscr{A}$. By Theorem 4.9, $\mathscr{A}^K = K$. Hence the element u_σ is determined up to a multiplier in E^*. Moreover, since $I_{u_\sigma} I_{u_\tau}$ and $I_{u_\sigma u_\tau}$ for σ, $\tau \in G$ have the same restriction $\sigma\tau$ to E we have $u_\sigma u_\tau = k_{\sigma,\tau} u_{\sigma\tau}$, $k_{\sigma,\tau} \in E^*$. Also $u_\sigma a u_\sigma^{-1} = \sigma a$, $a \in E$. Thus we have

$$u_\sigma a = (\sigma u) u_\sigma, \qquad u_\sigma u_\tau = k_{\sigma,\tau} u_{\sigma\tau} \tag{5.1}$$

for $a \in E$, $\sigma, \tau \in G$. The associativity $(u_\sigma u_\tau) u_\rho = u_\sigma (u_\tau u_\rho)$ gives the relations

$$k_{\sigma,\tau} k_{\sigma\tau,\rho} = k_{\sigma,\tau\rho}(\sigma k_{\tau,\rho}), \qquad \sigma, \tau, \rho \in G. \tag{5.2}$$

It is clear from (5.1) that the E-subspace $\sum_{\sigma \in G} E u_\sigma$ is a subalgebra. On the other hand, it is easily seen by a Dedekind independence argument that the u_σ are linearly independent over E. Hence $[\sum E u_\sigma : E] = |G| = n$ and hence $[\sum E u_\sigma : F] = n^2 = [A:F]$. Thus

$$\mathscr{A} = \sum E u_\sigma. \tag{5.3}$$

We now consider the converse in which we begin with the Galois extension field E/F and the Galois group G. Then a map k of $G \times G$ into E^*: $(\sigma, \tau) \rightsquigarrow k_{\sigma,\tau}$ is called a *Noether factor set* if (5.2) holds. We form the (left) vector space over E with base $\{u_\sigma | \sigma \in G\}$ and we define a product in $\mathscr{A} = \sum E u_\sigma$ by

$$\left(\sum a_\sigma u_\sigma\right)\left(\sum b_\tau u_\tau\right) = \sum_{\sigma,\tau} k_{\sigma,\tau} a_\sigma (\sigma b_\tau) u_{\sigma\tau}. \tag{5.4}$$

Then (5.2) implies that this is associative. Moreover, if we put $\sigma = \tau = 1$ and $\tau = \rho = 1$ successively in (5.2) we obtain

$$k_{1,\rho} = k_{1,1}, \qquad k_{\sigma,1} = \sigma k_{1,1}, \tag{5.5}$$

which imply that $1 = k_{11}^{-1} u_1$ is the unit of $\mathscr{A}$. Moreover, $\mathscr{A}$ is a vector space over $F \subset E$ and we have

$$\alpha(xy) = (\alpha x) y = x(\alpha y) \tag{5.6}$$

for $x, y \in \mathscr{A}$, $\alpha \in F$. Thus $\mathscr{A}$ is an algebra over F (associative with 1). This is called the *crossed product of E with G and Noether factor set k* and is denoted as $\mathscr{A} = (E, G, k)$. The result we proved above can now be stated as

5.7 Theorem. *If $\mathscr{A}$ is a central simple algebra containing E and the degree of $\mathscr{A}$ is $n = [E:F]$ then $\mathscr{A}$ is a crossed product (E, G, k).*

It is quite easy to prove directly the converse that any crossed product is central simple over F of degree $n = [E:F]$. We shall obtain this result by establishing the connection between crossed products and Brauer algebras. We note first that if we replace u_1 by 1 we may assume that the Noether factor set is normalized in the sense that $k_{1,\sigma} = 1 = k_{\sigma,1}$, $\sigma \in G$. Then we have

5.8 Theorem. *If k is a normalized Noether factor set then c defined by*

(5.9) $$c(\rho, \sigma, \tau) = \rho k_{\rho^{-1}\sigma, \sigma^{-1}\tau}$$

is a reduced Brauer factor set and $(E, G, k) \cong \mathscr{B}(E, c)$.

Conversely, if c is a reduced Brauer factor set then

(5.10) $$k_{\sigma,\tau} = c(1, \sigma, \sigma\tau)$$

defines a normalized Noether factor set and $\mathscr{B}(E, c) \cong (E, G, k)$.

Proof. For $a \in \mathscr{A} = (E, G, k)$ we write

(5.11) $$u_\sigma a = \sum_\tau a(\sigma, \tau) u_\tau,$$

where $a(\sigma, \tau) \in E$. Let $\mu(a)$ denote the matrix $(a(\sigma, \tau))$ (regarded as a map of $G \times G$ into E). Then $a \rightsquigarrow \mu(a)$ is a homomorphism of $\mathscr{A}$ into $M_n(E)$ since

$$u_\sigma ab = \textstyle\sum_\tau a(\sigma, \tau) u_\tau b = \sum_{\tau, \rho} a(\sigma, \tau) b(\tau, \rho) u_\rho$$

so $\mu(ab) = \mu(a)\mu(b)$. For $a \in E$, $\mu(a)$ is a diagonal matrix with σa in the (σ, σ)-position. Hence $E\mu(E) = \sum Ee_{ii}$. Also, by (5.1), if $v = \sum_{\sigma \in G} u_\sigma$ then $\mu(v) = (v(\sigma, \tau))$ where

(5.12) $$v(\sigma, \tau) = k_{\sigma, \sigma^{-1}\tau} \neq 0.$$

It follows as in the proof of Theorem 2.19 that $\mu(\mathscr{A})$ is central simple of degree n. Since $[\mathscr{A}:F] = n^2$, μ is a monomorphism and $\mathscr{A} = KvK$. Then the Brauer factor set determined by v is c where

$$c(\rho, \sigma, \tau) = v(\rho, \sigma)v(\sigma, \tau)v(\rho, \tau)^{-1} = k_{\rho, \rho^{-1}} k_{\sigma, \sigma^{-1}\tau} k^{-1}_{\rho, \rho^{-1}\tau} = \rho k_{\rho^{-1}\sigma, \sigma^{-1}\tau}$$

by the factor set condition (5.2). We have $c(\rho, \rho, \rho) = \rho k_{1,1} = 1$ so c is reduced. It is clear that $\mathscr{A} \cong \mathscr{B}(K, c)$.

Our result shows also that if K is a Noether factor set then c defined by (5.9) is a Brauer factor set. Direct verification shows that if c is a Brauer factor set then K defined by (5.10) is a Noether factor set and that the maps $k \rightsquigarrow c$ and $c \rightsquigarrow k$ are inverses. Now let $\mathscr{B}(E, c)$ be the Brauer algebra defined by c and let k be the corresponding Noether factor set. Consider the crossed product (E, G, k). Then since c is given by (5.9) the result we proved shows that $\mathscr{B}(E, c) \cong (E, G, k)$. □

The Noether factor sets form a group under multiplication which contains the subgroup of factor sets of the form

(5.13) $$(\sigma, \tau) \rightsquigarrow l_\sigma(\sigma l_\tau) l_{\sigma\tau}^{-1},$$

where $\sigma \rightsquigarrow l_\sigma$ is any map of G into E^*. The factor group is the cohomology group $H^2(G, E^*)$. As usual, we write $k \sim 1$ if k is in the subgroup defined by (5.13) and $k \sim k'$ if k and k' differ by an element of this subgroup.

We now observe that the map $k \rightsquigarrow c$ is an isomorphism of the group of Noether factor sets onto the group of Brauer factor sets. If l is any map of G into E^* then $l(\sigma, \tau) = \sigma l(\sigma^{-1}\tau)$ is a homogeneous map of $G \times G$ into E^*. It follows that if

$k \rightsquigarrow c$ in our homomorphism then $k \sim 1$ if and only if $c \sim 1$. Hence we have an induced isomorphism of $H^2(G, E^*)$ onto $H^2(E, F)$. This isomorphism together with Theorem 5.8 permit us to carry over the results of Section 4 to Noether factor sets and crossed products. We obtain in this way

5.14 Theorem. (i) *The crossed product* $(E, G, k) \sim 1$ *if and only if* $k \sim 1$.
(ii) $(E, G, k_1) \otimes_F (E, G, k_2) \sim (E, G, k_1 k_2)$.
(iii) *Let* $\{k\}$ *denote the element of* $H^2(G, E^*)$ *determined by* k. *Then* $\{k\} \rightsquigarrow \{(E, G, k)\}$ *is an isomorphism of* $H^2(G, E)$ *onto* $\mathrm{Br}(E/F)$. □

6. Central Simple Algebras with Involution of Orthogonal Type

In this section we assume char $F \neq 2$ (as well as F is infinite). Let $\mathscr{A}$ be a finite dimensional central simple algebra with an involution j (= anti-automorphism j such that $j^2 = 1_{\mathscr{A}}$). If $\bar{F}$ is the algebraic closure of F then j has a unique extension to an involution $\tilde{j}$ in $\tilde{\mathscr{A}} = \mathscr{A}_{\bar{F}} \cong M_n(\bar{F})$ and we can identify $\tilde{\mathscr{A}}$ with $M_n(\bar{F})$ and $\tilde{j}$ with either the transpose map $a \rightsquigarrow {}^t a$ or the map $a \rightsquigarrow s({}^t a)s^{-1}$ where

$$s = \mathrm{diag}\left\{\begin{pmatrix} 0 & 1 \\ -1 & 0 \end{pmatrix}, \begin{pmatrix} 0 & 1 \\ -1 & 0 \end{pmatrix}, \ldots, \begin{pmatrix} 0 & 1 \\ -1 & 0 \end{pmatrix}\right\}. \tag{6.1}$$

In the first case j is said to be of *orthogonal type* and in the second of *symplectic type*. These can be distinguished by the dimensionality of the space $\mathscr{H}(\mathscr{A}, j)$ of symmetric elements which is $n(n + 1)/2$ in the orthogonal case and $n(n - 1)/2$ (with n even) in the symplectic case. We can also distinguish the two cases by the fact that in the symplectic case if the degree of $\mathscr{A}$ is $2n$ then the degree of every element of $\mathscr{H}(\mathscr{A}, j)$ does not exceed n ([12], p. 231). On the other hand, we have

6.1 Theorem. *Let* $\mathscr{A}$ *be a central simple algebra of degree* n *over* F *with involution* j *of orthogonal type. Then* $\mathscr{H}(\mathscr{A}, j)$ *contains separable elements of degree* n *and for any such element* u *there exist* $v \in \mathscr{H}(\mathscr{A}, j)$ *such that* $\mathscr{A} = KvK$, $K = F[u]$.

Proof. As in Theorem 1.11, the subset of $\mathscr{H} = \mathscr{H}(\mathscr{A}, j)$ of separable elements of degree n is Zariski open in $\mathscr{H}$. Hence to show that it is not vacuous it suffices to show that the corresponding subset of $M_n(\bar{F})$ is not vacuous. This is the set of symmetric matrices having minimum polynomials with n distinct roots. Now $\sum r_i e_{ii}$ with distinct r_i is such a matrix. Now let u be any separable element of degree n in $\mathscr{H}$. If $v \in \mathscr{H}$ then $\mathscr{A} = F[u]vF[u]$ if and only if the elements $u^i v u^j$, $0 \le i, j \le n - 1$ are linearly independent. It is readily seen that the set of these v's is Zariski open in $\mathscr{H}$. Hence to prove that there exist such v's it suffices to show that the corresponding set in $M_n(\bar{F})$ is not vacuous. Now if $u \in M_n(\bar{F})$ is separable of degree n then there exists an orthogonal matrix s such that

$$sus^{-1} = \mathrm{diag}\{r_1, r_2, \ldots, r_n\}$$

where the r_i are distinct. Then we may take $v = \sum_{i,j} e_{ij}$, which is symmetric, to obtain n^2 linearly independent elements $u^i v u^j$. □

6.2 Remark. The argument we have used in the foregoing proof can be used to prove Theorem 2.2 for the case in which K is separable and F is infinite.

Now let K be a commutative separable algebra and let $\mathscr{B}(K, c)$ be a Brauer algebra defined by a reduced factor set c that is symmetric in the sense that

$$c(\alpha, \beta, \gamma) = c(\gamma, \beta, \alpha) \tag{6.3}$$

for $\alpha, \beta, \gamma \in M$. Then the map $l \rightsquigarrow {}^t l$ where

$${}^t l(\alpha, \beta) = l(\beta, \alpha) \tag{6.4}$$

is an involution of orthogonal type in $\mathscr{B}(K, c)$. For, we have $ll' = l''$ where

$$l''(\alpha, \beta) = \sum_{\gamma \in M} l(\alpha, \gamma)c(\alpha, \gamma, \beta)l'(\gamma, \beta)$$

and

$$\begin{aligned}({}^t l')({}^t l)(\alpha, \beta) &= \sum_{\gamma \in M} l'(\gamma, \alpha)c(\alpha, \gamma, \beta)l(\beta, \gamma)\\ &= \sum l(\beta, \gamma)c(\beta, \gamma, \alpha)l'(\alpha, \gamma) = l''(\beta, \alpha) = {}^t l''(\alpha, \beta).\end{aligned}$$

Hence $({}^t l')({}^t l) = {}^t l''$ and t is an involution. The subalgebra of $\mathscr{B}(K, c)$ consisting of the elements l such that $l(\alpha, \beta) = 0$ if $\alpha \neq \beta$ is isomorphic to K and is contained in $\mathscr{H}(\mathscr{B}(K, c), t)$. Since K contains an element of degree n it follows that t is of orthogonal type. We shall call t the *transpose involution* in $\mathscr{B}(K, c)$.

Two algebras with involution $(\mathscr{A}_1, j_1)$ and $(\mathscr{A}_2, j_2)$ are *isomorphic* if there exists an isomorphism n of $\mathscr{A}_1$ onto $\mathscr{A}_2$ such that $\eta j_1 = j_2 \eta$. With this definition we have

6.4 Theorem. *Let $\mathscr{A}$ be a central simple algebra with an involution j of orthogonal type. Then there exists a Brauer algebra $\mathscr{B}(K, c)$ with c symmetric such that $(\mathscr{A}, j)$ is isomorphic to $(\mathscr{B}(K, c), t)$.*

Proof. Let u and v be as in Theorem 6.1. Then every element of $\mathscr{A}$ can be written in one and only one way in the form $\sum_{k,l=1}^{n} a_{kl}u^{k-1}vu^{l-1}$. Since $u, v \in \mathscr{H}(\mathscr{A}, j)$, the involution j is

$$\begin{aligned}\sum_{k,l=1}^{n} a_{kl}u^{k-1}vu^{l-1} &\rightsquigarrow \sum a_{kl}u^{l-1}vu^{k-1}\\ &= \sum_{k,l} a_{lk}u^{k-1}vu^{l-1}.\end{aligned} \tag{6.5}$$

Now let E/F be a splitting field of $K = F[u]$ so $E = F(r_1, r_2, \ldots, r_n)$ where $f(\lambda) = \prod (\lambda - r_i)$ is the minimum polynomial of u over F. As in Section 2, we can take an imbedding of $\mathscr{A}$ in $M_n(E)$ so that $u = \operatorname{diag}\{r_1, r_2, \ldots, r_n\}$ and $v = (v_{ij})$ where every $v_{ij} \neq 0$. Then $\sum a_{kl}u^{k-1}vu^{l-1} = (l_{ij}v_{ij})$ where $l_{ij} = \sum a_{kl}r_i^{k-1}r_j^{l-1}$. Under our imbedding the involution j given by (6.5) becomes

$$(l_{ij}v_{ij}) = \sum a_{kl}u^{k-1}vu^{l-1} \rightsquigarrow \sum a_{lk}u^{k-1}vu^{l-1} = (l_{ji}v_{ij}). \tag{6.6}$$

Now we have the algebra $\mathscr{B}(K, c)$ which we regard, as in Section 2, as the set of matrices (l_{ij}) where the $l_{ij} \in E$ and $\sigma l_{ij} = l_{\sigma i, \sigma j}$ for $\sigma \in G = \text{Gal } E/F$. The multiplication in $\mathscr{B}(K, c)$ is $ll' = l''$ where $l = (l_{ij}), l' = (l'_{ij}), l'' = (l''_{ij})$, and $l''_{ij} = \sum_k l_{ik} c_{ikj} l'_{kj}$, $c_{ikj} = v_{ik} v_{kj} v_{ij}^{-1}$. Since $(l_{ij}) \rightsquigarrow (l_{ji})$ is an involution we have $\sum_k l'_{ki}(c_{ikj} - c_{jki}) l_{jk} = 0$ or

$$\sum_k l_{ik}(c_{ikj} - c_{jki}) l'_{kj} = 0 \tag{6.7}$$

for all l_{ik} and l'_{kj} satisfying the conjugacy conditions. As in the proof of (2.14) we can conclude that $c_{ikj} = c_{jki}$ for all i, k, j. If we now pass to the definition of $\mathscr{B}(K, c)$ as the algebra of homogeneous maps of $M \times M$ into E we see that this algebra has a symmetric Brauer factor set and the isomorphism $(l_{ij} v_{ij}) \rightsquigarrow$ is an isomorphism of $(\mathscr{A}, j)$ onto $(\mathscr{B}(K, c), t)$. □

References

[1] A. A. Albert. The structure of matrices with any normal division algebra of multiplications. *Ann. Math.*, **32** (1931), 131–148.

[2] A. A. Albert and H. Hasse. A determination of all normal division algebras over an algebraic number field. *Trans. Amer. Math. Soc.*, **34** (1932), 722–726.

[3] R. Brauer. Untersuchungen über die arithmetischen Eigenschaften von Gruppen linearer Substitutionen. *Math. Z.*, **8** (1928), 677–696.

[4] ———. On normal division algebras of degree 5. *Proc. Nat. Acad. Sci.*, **24** (1938), 243–246.

[5] R. Brauer, H. Hasse and E. Noether. Beweis eines Hauptaatzes in der Theorie der Algebren. *J. Math.*, **167** (1931), 399–404.

[6] R. Brauer and E. Noether. Uber minimale Zerfallungskorper irreduzibler Darstcllungen. *Sitzungsber. Preuss. Akad. Wiss.*, **32** (1927), 221–228.

[7] C. W. Curtis and I. Reiner. *Representation Theory of Finite Groups and Associative Algebras*. Interscience Publishers: New York, London.

[8] H. Hasse. Theory of cyclic algebras over an algebraic number field, *Trans. Amer. Math. Soc.*, **34** (1932), 171–214. Also Additional additional note to the authors: "Theory of cyclic algebras over an algebraic number field." *ibid*, 727–730.

[9] ———. Die Struktur der R. Brauerschen Algebren-Klassengresppen. *Math. Ann.*, **107** (1933), 731–760.

[10] N. Jacobson. *The Theory of Rings*. American Mathematical Society Surveys, 1943.

[11] N. Jacobson. Generation of separable and central simple algebras. *J. Math.*, **36** (1957), 217–227.

[12] ———. *Structure and Representations of Jordan Algebras*. Ann. Math. Soc. Colloquium Publ., 1968.

[13] ———. *Basic Algebra* II. W. H. Freeman and Company: San Francisco, 1980.

[14] E. Noether. *Gesamelte Abhandlungen*. Springer-Verlag: Berlin, Heidelberg, New York, 1982.

[15] H. Weyl. Generalized Riemann matrices and factor sets. *Ann. Math.*, **37** (1936), 709–745.

Noether's Problem in Galois Theory

RICHARD G. SWAN*

This paper is essentially an expanded version of [Sw] and gives a more detailed discussion of some of the topics mentioned there. It is intended to serve as an introduction to recent work related to Noether's problem. To avoid too much overlap with [Sw] a number of historical remarks, comments, and other topics have been omitted here.

1. Introduction

Noether's work on Galois theory was concerned with the following questions.

(1) Can any finite group G be realized as a Galois group over a given number field F?
(2) If so, can all such extensions be parametrized in a simple way?

References to recent work on (1) are given in [Sw]. In [11], Noether showed that both questions have an affirmative answer if the following question has.

(3) Let G act faithfully by permutations on a finite set of indeterminates $x_1, \ldots, x_n$. Is the fixed field $F(x_1, \ldots, x_n)^G$ a purely transcendental extension of F?

This question is generally referred to as Noether's problem. Although the answer to (3) is negative in general, it has been the inspiration for a great deal of recent work in a number of seemingly unrelated areas. I will describe some of these developments here beginning with Saltman's beautiful work on generic Galois extensions which is very closely related to Noether's paper ([11]) and gives, in particular, a very elegant formulation of her results.

* Department of Mathematics, University of Chicago, 5734 University Avenue, Chicago, IL 60637, U.S.A.

Remark. In [Sw] I asked why Noether's name was associated with the cocycle relations used in the non-cyclic generalization of Hilbert's Theorem 90. N. Jacobson remarked that this was undoubtedly due to Noether's pioneering work in the application of cohomology to number theory ([42]). Fröhlich [Fr] gives a very nice discussion of this using modern notation.

2. Galois Extensions

The definition of a Galois extension of rings was slightly misstated in [Sw]. The correct definition is as follows. Let $A \subset B$ be commutative rings. Let G be a finite group of automorphisms of B. Then B is a Galois extension of A with group G if

(1) $A = B^G$.
(2) For all subgroups $H \subset G$ and all H-stable ideals I of B with $I \neq B$, H acts faithfully on B/I.

In [Sw], (2) was only stated for the case $H = G$. The Galois theory of commutative rings is developed in [CHR] and [DMI]. In this section I will give a short exposition of the properties we will need.

Proposition 2.1. *Let G be a finite group of automorphisms of a commutative ring B and set $A = B^G$. The following are equivalent:*

(1) *B is a Galois extension of $A = B^G$ with group G.*
(2) *There are $x_i, y_i \in B$ such that $\sum x_i\sigma(y_i) = \delta_{\sigma 1}$ for all $\sigma \in G$. Here $\delta_{\sigma\tau}$ is 1 if $\sigma = \tau$ and 0 otherwise.*
(3) *$h: B \otimes_A B \xrightarrow{\approx} \prod_{\sigma \in G} B$ where $h(b_1 \otimes b_2)$ has coordinates $h_\sigma(b_1 \otimes b_2) = b_1\sigma(b_2)$.*

Proof. If (B, G) satisfies (2) so does (B, H) for $H \subset G$ and so does $(B/I, H)$ if I is H-stable. Therefore (2) clearly implies (1). It is trivial that (3) implies (2) since (2) is equivalent to the existence of $z = \sum x_i \otimes y_i \in B \otimes_A B$ with $h(z) = (1, 0, \ldots, 0)$.

Suppose that (1) holds. If $\sigma \neq 1$ in G then σ acts trivially on B/I where I is the ideal generated by all $\sigma(b) - b$, $b \in B$. Therefore $I = B$ so $\sum a_i(\sigma b_i - b_i) = 1$ for some $a_i, b_i \in B$, $i = 1, \ldots, n$. Let $a_0 = -\sum a_i\sigma(b_i)$ and $b_0 = 1$. Then $z_\sigma = \sum a_i \otimes b_i$ has $h_1(z_\sigma) = 1$, $h_\sigma(z_\sigma) = 0$ so $z = \prod_{\sigma \neq 1} z_\sigma$ has $h(z) = (1, 0, \ldots, 0)$ which is (2).

If (2) holds, let

$$k: \prod B \to B \otimes_A B \quad \text{by} \quad k((b_\sigma)) = \textstyle\sum_\sigma \sum_i x_i \otimes \sigma^{-1}(y_i b_\sigma).$$

Then $hk = id$ and

$$\begin{aligned} kh(s \otimes t) &= \textstyle\sum\sum x_i \otimes \sigma^{-1}(s\sigma(t)y_i) = (1 \otimes t)\sum x_i \otimes \operatorname{tr}(sy_i) \\ &= (1 \otimes t)\textstyle\sum x_i \operatorname{tr}(sy_i) \otimes 1 = (1 \otimes t)\sum\sum x_i\sigma(y_i)\sigma(s) \otimes 1 \\ &= (1 \otimes t)(s \otimes 1) = s \otimes t \end{aligned}$$

since $\operatorname{tr}(b) = \sum \sigma(b) \in A$. □

Corollary 2.2. *If B is Galois over A then B is finitely generated and projective as an A-module.*

In fact, the identity map of B factors as $B \to A^n \to B$ by $b \to (\mathrm{tr}(y_i b))$ and $(a_i) \to \sum a_i x_i$.

Proposition 2.3. *If B is Galois over A then* $\mathrm{tr}: B \to A$ *is onto.*

Proof. Let $I = \mathrm{tr}(B)$. It is an ideal of A and $IB = B$ since $b = \sum x_i \, \mathrm{tr}(y_i b)$ as in Corollary 2.2. By [Ka, Th. 76] we can find $r \in I$ with $(1 - r)B = 0$. Since $1 \in B$, $r = 1$ and $I = A$. □

Proposition 2.4. *If B is a Galois extension of A with group G and $A \to C$ is any homomorphism of commutative rings then $C \otimes_A B$ is a Galois extension of C with group G.*

Proof. $C \otimes_A B$ satisfies (2) of Proposition 2.1 using $1 \otimes x_i$ and $1 \otimes y_i$. Since the monomorphism $A \to B$ is split by $b \to \mathrm{tr}(bc)$ where $c \in B$ has $\mathrm{tr}(c) = 1$, we see that $C \to C \otimes_A B$ is injective. By Proposition 2.3,

$$(C \otimes_A B)^G = \mathrm{tr}(C \otimes_A B) = C \otimes_A \mathrm{tr}(B) = C$$

since tr is A-linear. □

It is clear that $\prod_{\sigma \in G} A$ with G action by $\sigma((x_\tau)) = (x_{\tau\sigma})$ is a Galois extension of A with group G. We can choose $x_\sigma = y_\sigma = e_\sigma = (\delta_{\sigma\tau})$ in Proposition 2.1(2). A split Galois extension is one isomorphic to $\prod_G A$ preserving the G action. For any Galois extension $A \subset B$, we can find a faithfully flat extension C of A such that $C \otimes_A B$ splits. In fact, $C = B$ will do by Proposition 2.1(3).

Using this remark, it is easy to generalize the classical descent theorem for vector spaces ([Sp]).

Proposition 2.5. *Let $A \subset B$ be a Galois extension with group G. Let M be a B-module with G-action such that $\sigma(bm) = \sigma(b)\sigma(m)$ for $b \in B$, $m \in M$. Then $B \otimes_A M^G \xrightarrow{\approx} M$ by $b \otimes m \to bm$.*

Proof. All the data are preserved by a faithfully flat base extension so we can assume that B splits. Therefore, $M = \prod M_\sigma$ with $M_\sigma = e_\sigma M$, e_σ being as above. All the M_σ are isomorphic as A-modules since $\sigma: M_1 \approx M_\sigma$. It is now clear that $M^G \approx M_1$ and $B \otimes_A M^G \xrightarrow{\approx} M$. □

Suppose B is a Galois extension of A with group H and $H \subset G$. We can then define an induced Galois extension $\mathrm{Ind}_H^G(B)$ of A with group G. Let $B' = \mathrm{Ind}_H^G(B)$ be the set of functions $f: G \to B$ such that $f(hg) = hf(g)$ for $h \in H$, $g \in G$. Then B' is a ring under pointwise addition and multiplication and $B' \approx \prod_{H \backslash G} B$ as a ring. Let G act on B' by $\sigma f \cdot (\tau) = f(\tau\sigma)$. One checks easily that $B'^G = A$. Let $G = \bigcup H\tau_k$. If $x_i, y_i \in B$ are as in Proposition 2.1(2), let $f_{ij}(\tau_k)$ be 0 for $j \neq k$ and x_i for $j = k$, and similarly, $g_{ij}(\tau_k) = \delta_{jk} y_i$. Then $\sum_{i,j} f_{ij}\sigma(g_{ij}) = \delta_{\sigma 1}$ so B' is Galois over A.

Proposition 2.6. *Let B be a Galois extension of a field F with group G. Then $B \approx \mathrm{Ind}_H^G(K)$ where K is a Galois field extension of F.*

Proof. If $I = \mathrm{rad}(B) \neq 0$, find n so $J = I^n \neq 0$ but $J^2 = 0$. Clearly J is G-stable and $J \otimes J \neq 0$ but $h(J \otimes J) = 0$ contradicting Proposition 2.1(3). Therefore B is semisimple so $B = B_1 \times \cdots \times B_m$ where the B_i are fields. G permutes the B_i transitively since otherwise $B = B' \times B''$ as a G-module so $F = B'^G \times B''^G$. Let H be the stabilizer of $K = B_1$. It is clear that $B \approx \mathrm{Ind}_H^G(K)$. □

3. Transplanting Galois Extensions

Suppose K/F is a Galois extension of fields with group G. We want to construct other Galois extensions with group G by "lifting" the given extension to a subring A of F and applying Proposition 2.4.

Lemma 3.1. *Let K/F be a Galois extension of fields with group G. Let $B \subset K$ be a subring with quotient field K which is stable under G and set $A = B^G$. Then we can find $s \in A$, $s \neq 0$, such that $A[s^{-1}] \subset B[s^{-}1]$ is a Galois extension.*

Proof. Clearly b/s^n is fixed under G if and only if b is so $B[s^{-1}]^G = A[s^{-1}]$. Let $x_i, y_i \in K$ be as in Proposition 2.1(2). Find $t \in B$ such that $x_i, y_i \in B[t^{-1}]$ and set $s = \prod_{\sigma \in G} \sigma(t)$. □

Remark. The last step in this argument shows that any element of K can be written as b/a with $a \in A$. It follows that F is the quotient field of A.

By replacing A, B by $A[s^{-1}]$, $B[s^{-1}]$ we can now assume that $A \subset B$ is a Galois extension. Let $A \to E$ be a homomorphism with E a field. By forming $E \otimes_A B$ we get a Galois extension of E. However, it may only be a product of fields as in Proposition 2.6. We must, therefore, seek conditions under which $E \otimes_A B$ will be a field.

Let $K = F[\theta]$. By using $a\theta$ in place of θ (with $a \in A$), we can assume that $\theta \in B$. Let f be the minimal polynomial of θ so $f(T) = \prod_{\sigma \in G} (T - \sigma(\theta))$. Let $A \to E$ send f to $g(T) \in E[T]$. If g is irreducible then $1 \otimes \theta \in E \otimes_A B$ generates a field $\approx E[T]/(g(T))$ of degree $n = \deg g = \deg f = |K:F|$ over E. But $n = |E \otimes_A B : E|$. This follows easily from Corollary 2.2 and the fact that projective modules over local rings are free ([B, III, 2.13]) since we can replace A by A_P where $P = \ker[A \to E]$ and B by B_{A-P} which is free over A_P. It follows that $E \otimes_A B = E[T]/(g(T))$ which is a field. Therefore it will suffice to find $A \to E$ such that the image g of f is irreducible over E. In some cases this can be done using Hilbert's irreducibility theorem:

Theorem 3.2. *Let $F = k(t_1, \ldots, t_r)$ be a rational (=pure transcendental) extension of a number field k. Let $f(T) = f(t_1, \ldots, t_r, T) \in F[T]$ be irreducible over F. Then there are $a_1, \ldots, a_r \in k$ such that $f(a_1, \ldots, a_r, T)$ is defined and irreducible over k.*

For a simple proof (and a more general statement) see [LDG].

In order to apply this we must assume that $F = k(t_1, \ldots, t_r)$ as in Theorem 3.2. Then K is finitely generated over k so we can choose B to be of finite type over k. It is well known ([AT1]) that the same will then be true of A. In fact, let $A' \subset A$ be generated by the coefficients of the polynomials $\prod (X - \sigma(\xi))$ where ξ runs over a set of generators for B over k. Then B and hence A is finite over A'. It follows that we can find $h \in k[t_1, \ldots, t_r]$ with $A \subset k[t_1, \ldots, t_r, h^{-1}]$. Let $f(T)$ be as above and apply Theorem 3.2 to f/h getting $a_1, \ldots, a_r \in k$ such that $h(a_1, \ldots, a_r) \neq 0$ and such that the image g of f under $k[t_1, \ldots, t_r, h^{-1}] \to E = k$ by $t_i \to a_i$ is irreducible. The above argument then shows that $k \otimes_A B$ is a Galois field extension with group G.

We can now derive Noether's first result on realizing Galois groups. If G is any finite group we can find a faithful permutation representation on a finite set $x_1, \ldots, x_m$ of indeterminates. G then acts on $k(x_1, \ldots, x_m)$.

Theorem 3.3. (Noether). *Let k be a number field. If $k(x_1, \ldots, x_m)^G$ is a rational extension of k, there is a Galois field extension L/k with group G.*

4. Generic Galois Extensions

We now discuss Noether's idea of parametrizing all Galois extensions with a given group ([11]), following Saltman [Sa].

Definition 4.1 (Saltman). *A Galois extension $A \subset B$ with group G is called a generic Galois extension for the field k and group G if:*

(1) *$A = k[x_1, \ldots, x_n, s^{-1}]$ where the x_i are indeterminates and $s \in k[x_1, \ldots, x_n]$.*
(2) *If F is a field containing k and K is a Galois extension of F with group G, there is a k-algebra homomorphism $A \to F$ such that $F \otimes_k B \approx K$ as a Galois extension, i.e. preserving the G action.*

In other words, all Galois extensions of fields $F \supset k$ with group G can be parametrized by maps $A \to F$ or, equivalently, by the images of the x_i in F.

Note that K in (2) is not required to be a field. See Remark 4.6.

Noether's result on parametrizing Galois extensions can now be expressed as follows ([Sa]).

Theorem 4.2. *Let G act faithfully on a finite set of indeterminates $x_1, \ldots, x_n$. Let k be an infinite field. If $k(x_1, \ldots, x_n)^G$ is rational over k then there is a generic Galois extension for k and G.*

In Noether's original formulation ([11]) certain exceptional cases were excluded. There was little point in worrying about these at the time since the essential question was whether the hypothesis of rationality holds. The exceptions were later eliminated by Kuyk [Ku].

The theorem is proved by the method of Section 3. We start with $B_0 = k[x_1, \ldots, x_n]$ and $A_0 = B_0^G$ and pass to a Galois extension

$$A = A_0[s^{-1}] \subset B_0[s^{-1}] = B.$$

This will be the required generic extension for suitable s. The rationality of $k(x_1, \ldots, x_n)^G = k(y_1, \ldots, y_n)$ is only needed to get condition (1).

Lemma 4.3. *Let C, D be k-algebras of finite type with the same quotient field. Then there are elements $c \in C$, $d \in D$ such that $C[c^{-1}] = D[d^{-1}]$ (with $c, d \neq 0$).*

Proof. If $c_1, \ldots, c_r$ generate C over k find $a \in D$ so that all c_i lie in $D[1/a]$. Then $C \subset D[1/a]$. The same argument now gives $D[1/a] \subset C[1/c]$ so $C[1/c] = D[1/ab]$ where $c = b/a^m$.

This shows that we can assume that $A = k[y_1, \ldots, y_n, s^{-1}]$. We now turn to the proof of (2) which does not require the rationality assumption.

Lemma 4.4. *Let $A \subset B$ and $C \subset D$ be Galois extensions with the same group G. Let $f: B \to D$ be a G-equivariant homomorphism. Then f induces an isomorphism $C \otimes_A B \xrightarrow{\approx} D$.*

Proof. We can replace $A \subset B$ by $C \subset C \otimes_A B$ and thus assume that f induces the identity on $A = C$. Define $\langle x, y \rangle = \mathrm{tr}(xy)$ for $x, y \in B$. This is a dual pairing, i.e. it induces an isomorphism $B \xrightarrow{\approx} \mathrm{Hom}_A(B, A)$. It is sufficient to check this after making a faithfully flat extension of A but it is obvious in the split case. Clearly

$$\langle f(x), f(y) \rangle = \langle x, y \rangle$$

since $f | A = \mathrm{id}$. If $f(x) = 0$, it follows that $\langle x, B \rangle = 0$ so $x = 0$. Therefore f is injective. If A is a field (the only case needed here) f is an isomorphism by comparing dimensions. To show f is onto in general we can localize A to A_P for all prime ideals P of A and then pass to A_P/PA_P by Nakayama's lemma. This reduces the question to the field case. □

The following lemma of Kuyk [Ku] now provides the required equivariant map $B \to K$.

Lemma 4.5 (Kuyk). *Let G act on a finite set $X = \{x_1, \ldots, x_n\}$. Let $F \subset K$ be a Galois extension with group G such that F is an infinite field. Let $s \in F[x_1, \ldots, x_n]$, $s \neq 0$. Then there is a G-equivariant map $f: X \to K$ such that $s(f(x_1), \ldots, f(x_n)) \in K^*$.*

Proof. Let $x_1, \ldots, x_m$ be a set of representatives for the orbits of G on X. Let H_i be the stabilizer of x_i. Set $f(x_i) = a_i$ where $a_i \in K^{H_i}$ and extend by $f(gx_i) = ga_i$ for $g \in G$. We must choose the a_i so that $s(f(x)) \in K^*$. By Proposition 2.6, $K = \mathrm{Ind}_H^G(L)$ with L a field. We must find the a_i so all coordinates of $s(f(x))$ in $K = L \times \cdots \times L$ are non-zero. Let w_{ij} be a base for K^{H_i}. Let $a_i(Z) = \sum_j Z_{ij} w_{ij}$. The coordinates of $s(a(Z))$ are polynomials in the Z_{ij}. It is enough to show that these polynomials are non-zero. Then, since F is infinite, we can specialize the Z_{ij} to values in F such that the polynomials do not vanish. The required non-vanishing is clearly not affected by a change of base or an extension of the ground field F. By extending F to L we can assume that K splits since Ind_H^G clearly commutes with extension of F. As a base for K^{H_i} we can now choose the elements $\sum_{\tau \in H_i} e_{\sigma\tau}$ where e_σ is as in Section 2. Therefore $a_i = (Z_{i\sigma}) \in \prod_\sigma F$ where $Z_{i\sigma} = Z_{i\sigma\tau}$ for $\tau \in H_i$. The σth coordinate of $s(a)$ is $s(Z_{1\sigma}, \ldots, Z_{n\sigma})$ and is obviously non-zero. □

Remark 4.6. As a consequence of the above arguments it is easy to see that if k is infinite it is sufficient to assume condition (2) of Definition 4.1 for the case where K is a field.

Suppose that $R \subset S$ satisfies this modified definition. Then we can find an equivariant map $g: S \to k(x_1, \ldots, x_n)$. By a suitable choice of s we can assume that $g(S) \subset B$. Now if K is any Galois extension of F we get the required equivariant map as the composition $S \to B \to K$.

Note that this may fail if k is finite. For example, let G be abelian non-cyclic of order prime to the characteristic of k. By Theorem 5.5 there is a generic Galois extension $A \subset B$ for k and G. Let $A' = A \otimes_k k[x, 1/f(x)]$ and similarly for B' where $f(a) = 0$ for all $a \in k$. Then $A' \subset B'$ satisfies the modified condition (2) which is now vacuous for F finite, but there is no map $A' \to k$.

5. Grunwald's Theorem

Grunwald's theorem in classfield theory asserted the existence of cyclic extensions of number fields with prescribed local properties at a finite number of primes. It was accepted as correct for nearly 20 years until Wang [W1] found a counter-example.

Theorem 5.1 (Wang). *If $8/n$, there is no cyclic field extension $K/\mathbf{Q}$ such that $K \otimes_{\mathbf{Q}} \mathbf{Q}_2$ is an unramified cyclic field extension of $\mathbf{Q}_2$ of degree n.*

A proof is given at the end of this section. In ⟨W2⟩, Wang gave a correct version of the theorem. An account of the Grunwald–Wang theorem is given in [AT].

Saltman [Sa] discovered a completely unexpected relation between Grunwald's theorem and Noether's theorem 4.2. Using this he showed that Theorem 5.1 implies the falsity of Noether's question over $\mathbf{Q}$ for the case of a cyclic group of order 8. This case had, in fact, been found previously by Lenstra [L] and Voskresenskiĭ [V] but the connection with Grunwald's theorem came as a complete surprise. Saltman's approach is based on the following result which could be thought of as an analogue of the homotopy property for pullbacks of fiber bundles. Suppose $F \supset k$ is a local field and $A = k[x_1, \ldots, x_n, s^{-1}]$. We say that $g: A \to F$ is near $f: A \to F$ if $g(x_i)$ is near $f(x_i)$ for all i in the usual topology of F.

Lemma 5.2. *If $A \subset B$ is a Galois extension and $g: A \to F$ is sufficiently near $f: A \to F$ then $F \otimes_{A,g} B \approx F \otimes_{A,f} B$ as Galois extensions.*

Proof. I will only discuss the case where $K = F \otimes_{A,f} B$ is a field. For the general case see [Sa, pp. 281–282]. Replace A by $F \otimes_k A$ and B by $F \otimes_k B$ so that $B \to F \otimes_{A,f} B = K$ is onto. Let $K = F(c)$ and write c as $1 \otimes b$, $b \in B$. Let

$$\Phi(X) = \prod (X - \sigma b).$$

Then Φ_f, the result of applying f to the coefficients of Φ, is the minimal polynomial of c. If g is near f, Φ_g is near Φ_f and so is irreducible ([A, Ch. 2, §6, Th. 9]). Since $F \otimes_{A,g} B$ contains a root $1 \otimes b$ of Φ_g it must, by comparing dimensions, be the field obtained from F by adjoining such a root. This is isomorphic to K by [A, Ch. 2, §6, Th. 9] again. □

Corollary 5.3 (Saltman). *Let F be an algebraic number field such that there is a generic Galois extension for F and G. Let v be a valuation of F and let L be a Galois extension of the completion F_v with group G. Then there is a Galois extension K of F with group G so that $F_v \otimes_F K \approx L$.*

We could even specify the local extensions L_i of F_{v_i} at any finite set of valuations v_i.

Proof. Let $A \subset B$ be the generic extension. Let $f: A \to F_v$ induce L. Choose $b_i \in F$ very near $a_i = f(x_i)$ and let $g(x_i) = b_i$. This induces the required K. It is clear from this argument why the algebraic independence of the generators x_i in Definition 4.1 is essential. □

Corollary 5.4. *If G is cyclic of order $n \equiv 0 \pmod 8$ and G acts faithfully on a set of indeterminates $x_1, \ldots, x_n$ then $\mathbf{Q}(x_1, \ldots, x_n)^G$ is not rational over $\mathbf{Q}$.*

If it was, Theorem 4.2 would imply the existence of a generic Galois extension for $\mathbf{Q}$ and G. This is impossible by Corollary 5.3 and Theorem 5.1.

In Wang's treatment of Grunwald's theorem the question of the cyclicity of $k(\zeta)/k$ for certain 2^mth roots of unity ζ plays a key role. Saltman has shown that this too is intimately connected with the problem of generic Galois extensions. The following is one of the main results of [Sa].

Theorem 5.5 (Saltman). *Let G be an abelian group of order prime to* char k. *Let r be the largest power of 2 dividing the exponent of G. If $k(\zeta_r)/k$ is cyclic, there is a generic Galois extension for k and G.*

If k is a number field, it follows from Corollary 5.3 that Grunwald's theorem holds in its strongest form if the hypothesis of Theorem 5.5 is satisfied, i.e. given Galois extensions L_v of k_v with group G for a finite number of valuations v, there is a Galois extension K of k which localizes to L_v at v for each v.

By using this and some refinements in the non-cyclic case, Saltman was able to give an elementary version of the Grunwald–Wang theorem which is adequate for the classical application: the theorem that all simple algebras over a number field are cyclic.

Wang's proof of Theorem 5.1 used classfield theory. Here is an elementary proof along the same lines. Suppose there is an extension $K/\mathbf{Q}$ with the stated property. By using the subfield of K fixed by G_{odd} we can assume that n is a power of 2. Let $K \supset L \supset \mathbf{Q}$ with $|L:\mathbf{Q}| = 2$. Since 2 does not split or ramify in K, the same is true in L so $L = \mathbf{Q}(\sqrt{m})$ where $m \equiv 5 \pmod 8$. Let $p|m$ be a prime with $p \not\equiv 1 \pmod 8$. Then p ramifies in L. The inertia group for p in $\text{Gal}(L/\mathbf{Q})$ is the image of the inertia group T of p in G. Since $G = \mathbf{Z}/n\mathbf{Z}$, $n = 2^a$, we see that $T = G$, i.e. p is totally ramified in K. All this is taken directly from [W1]. The following lemma now gives the required contradiction $n|p - 1$.

Lemma 5.6. *Let K/F be a Galois extension of number fields or local fields with group G cyclic of prime power order n. Let $\mathfrak{p}$ be a prime of F totally ramified in K and such that $\mathfrak{p} \nmid n$. Then n divides $N\mathfrak{p} - 1$.*

Proof. By completing at $\mathfrak{p}$ it is sufficient to do the local case. Let v be the valuation. From $1 \to U \to K^* \xrightarrow{v} \mathbf{Z} \to 0$ we get $H^0(G, K^*) \to H^0(G, \mathbf{Z}) \to H^1(G, U)$ or $F^* \xrightarrow{v} \mathbf{Z} \to H^1(G, U)$. But $v(F^*) = n\mathbf{Z}$ so $\mathbf{Z}/n\mathbf{Z}$ embeds in $H^1(G, U)$. Now consider $1 \to U_1 \to U \to k^* \to 1$ where k is the residue field. Since U_1 is an inverse limit of finite p-groups, $\mathfrak{p}|p$, we see that U_1 is uniquely divisible by n so $0 = H^1(G, U_1) \to H^1(G, U) \to H^1(G, k^*)$. Therefore $\mathbf{Z}/n\mathbf{Z}$ embeds in $H^1(G, k^*) = \text{Hom}(G, k^*) \subset k^*$ so n divides $|k^*| = N\mathfrak{p} - 1$. □

6. Fischer's Theorem

Noether's problem is still of great interest in the case where the ground field is algebraically closed even though there is no direct application to Galois theory in this case. The case where G is abelian was solved by E. Fischer [F]. Fischer, who replaced Gordon at Erlangen, had a considerable influence on Noether [Ki]. According to [5], her work on Galois theory originated in conversations with Fischer. Fischer's solution has since been rediscovered a number of times ([L]). The result can be stated in somewhat greater generality by observing that $k[x_1, \ldots, x_n]$ is the symmetric algebra of the vector space $\sum kx_i$ [Ke].

Theorem 6.1 (Fischer). *Let G be a finite abelian group of exponent e. Let k be a field of characteristic prime to e and containing the eth roots of unity. Let V be a finite dimensional representation of G over k and let $k(V)$ be the quotient field of the symmetric algebra $S_k(V)$. Then $k(V)^G$ is rational over k.*

Proof. Write $V = \sum V_i$ where $V_i = kx_i$ is 1-dimensional and G acts by $\sigma x_i = \chi_i(\sigma)x_i$ where χ_i is a character of G. Let $X \subset k(V)^*$ be the multiplicative group generated by the x_i. Then $k(V)$ is the quotient field of $k[x_1, x_1^{-1}, \ldots, x_n, x_n^{-1}] = k[X]$, the group ring of the free abelian group X over k. Let $\hat{G} = \text{Hom}(G, k^*)$ be the character group of G and define $X \to \hat{G}$ by sending x_i to χ_i. If Y is the kernel of this map, one verifies trivially that $k[X]^G = k[Y]$. Since Y is free abelian, $k(V)^G = k(Y)$ is rational over k. □

Some interesting results on the metabelian case have recently been obtained by M. Hajja [H].

7. Galois Descent

Masuda had the idea of attacking Noether's problem by extending the ground field, applying Fischer's theorem, and using Galois descent to get back to the original field ([M]). The rest of this paper will be devoted to work based on this key idea.

Let G be abelian of exponent e prime to the characteristic of k and set $k' = k(\zeta_e)$ with ζ_e a primitive eth root of unity. Let $\pi = \mathrm{Gal}(k'/k)$. If V is a finite dimensional representation of G over k, we can apply Fischer's theorem to $k' \otimes_k k(V)$. A bit more care is required to keep track of the action of π. Let $V' = k' \otimes_k V = \bigoplus V_\chi$ where χ runs over the character group $\hat{G} = \mathrm{Hom}(G, k'^*)$ and V_χ is the set of $v \in V'$ with $\sigma v = \chi(\sigma)v$ for $\sigma \in G$. If $\alpha \in \pi$, $\alpha V_\chi = V_{\alpha(\chi)}$ so π permutes the V_χ. Let π_χ be the stabilizer of V_χ. By Proposition 2.5, V_χ has a base fixed by π_χ. Applying $\alpha \in \pi$ to this gives a base for $V_{\alpha\chi}$. In this way we find a base $x_1, \ldots, x_n$ for V' which is permuted by π and is such that each x_i lies in some V_χ. The proof of Theorem 6.1 can now be applied. The group X will be a permutation module over $\mathbf{Z}\pi$, i.e. it has a base permuted by π. The map $X \to \hat{G}$ is clearly π-equivariant where π acts on G by letting α send χ to $\alpha(\chi)$. Therefore, the kernel Y will be a $\mathbf{Z}\pi$-module. Since $k' \otimes_k k(V)^G = (k' \otimes_k k(V))^G = k'(Y)$, we have $k(V)^G = k'(Y)^\pi$. This leads to the following construction:

Let k'/k be a Galois field extension with group π. Let M be a $\mathbf{Z}\pi$-module which is free and finitely generated over $\mathbf{Z}$. Form the group ring $k'[M]$ with π acting on both k' and M. Define $A(k'/k, M) = k'[M]^\pi$ and let $Q(k'/k, M) = k'(M)^\pi$ be the quotient field of $A(k'/k, M)$. Note that $k'[M] = k' \otimes_k A(k'/k, M)$ and $k'(M) = k' \otimes_k Q(k'/k, M)$ by Proposition 2.5.

Lemma 7.1 (Masuda). *If M is a permutation module then $Q(k'/k, M)$ is rational over k.*

Proof. Let M have base $x_1, \ldots, x_n$ permuted by π. Then $W = \sum k'x_i \subset k'(M)$ generates $k'(M)$. By Proposition 2.5, $W = k' \otimes_k W^\pi$ which is easily seen to imply $k'(W)^\pi = k(W^\pi)$. □

Masuda used his method to show that $k(x_1, \ldots, x_n)^G$ is rational for certain groups G. In [SwI] I gave a partial converse to Masuda's method which showed that $\mathbf{Q}(x_1, \ldots, x_{47})^G$ for $G = \mathbf{Z}/47\mathbf{Z}$ is not rational. Voskresenskiĭ [V] found a sharper form of the invariant.

Let K/k be a finitely generated field extension with k algebraically closed in K. If k'/k is a finite Galois extension with group π, $k' \otimes_k K$ will again be a field. We look for a finitely generated k'-subalgebra $A \subset k' \otimes_k K = K'$ with quotient field K' and stable under π. The required invariant is defined by using the π-module A^*/k'^*. We must see to what extent this will be independent of the choice of A. If B is another such ring we can find $a \in A^\pi$ and $b \in B^\pi$ with $A[a^{-1}] = B[b^{-1}]$ by the proof of Lemma 4.3. Therefore it is enough to compare A with $A[a^{-1}]$. Since $a \in A^\pi$, the prime factors $p_1, \ldots, p_t$ of a will be permuted by π up to units. Therefore there is an exact sequence of $\mathbf{Z}\pi$-modules $0 \to A^* \to A[a^{-1}]^* \to S \to 0$ where S is the permutation module of formal $\mathbf{Z}$-linear combinations of $(p_1), \ldots, (p_t)$ and $A[a^{-1}]^* \to S$ by $x \to \sum \operatorname{ord}_{p_i}(x)(p_i)$. This suggests the following definition.

Definition 7.2. Let F_π be the set of equivalence classes of finitely generated torsion free $\mathbf{Z}\pi$-modules under the equivalence relation generated by $M \sim N$ if there is a short exact sequence $0 \to M \to N \to S \to 0$ where S is a permutation module. Write $\rho(M)$ for the class of M in F_π. If K is as above and A^*/k'^* is finitely generated write $\rho(K)$ for the class of A^*/k'^* in F.

Note that F_π is clearly a monoid under direct sum. An alternative description is given in Section 8. The above argument shows that $\rho(K)$ is a birational invariant of K/k. Taking $A = k'[M]$ shows that $\rho(Q(k'/k, M)) = \rho(M)$.

8. $\mathbf{Z}\pi$-modules

The invariant used in [SwI] was the image of $\rho(K)$ in the Grothendieck group $G_0(\mathbf{Z}\pi)$ mod permutation modules. However, in subsequent investigations it was necessary to use the sharper form of the invariant given in Section 7. The algebra required to deal with this was developed by Voskresenskiĭ, Endo and Miyata, Lenstra, and Colliot-Thélène and Sansuc. The account given here follows the first section of [CTS].

Let $\mathscr{L}_\pi$ be the category of $\mathbf{Z}\pi$-modules which are free and finitely generated over $\mathbf{Z}$. Let Perm_π be the full subcategory of permutation modules and let $\mathscr{P}_\pi$ be the full subcategory of modules which are direct summands of permutation modules. These are referred to as permutation projective in [L] and as invertible in [CTS].

Let $\mathscr{F}_\pi$ be the full subcategory of $\mathscr{L}_\pi$ of modules F such that all short exact sequences $0 \to P \to E \to F \to 0$ with $P \in \mathscr{P}_\pi$ split, i.e. $\operatorname{Ext}^1_{\mathbf{Z}\pi}(F, P) = 0$ for $P \in \mathscr{P}_\pi$. These modules are called flasque in [CTS].

Let $\mathscr{C}_\pi$ be the full subcategory of modules C such that all short exact sequences $0 \to C \to E \to P \to 0$ with $P \in \mathscr{P}_\pi$ split, i.e. $\operatorname{Ext}^1_{\mathbf{Z}\pi}(P, C) = 0$ for $P \in \mathscr{P}_\pi$. These modules are called coflasque in [CTS].

The duality functor $M \mapsto M^\vee = \operatorname{Hom}_{\mathbf{Z}}(M, \mathbf{Z})$ gives an involution of $\mathscr{L}_\pi$ which clearly preserves permutation modules. Therefore $\mathscr{P}_\pi$ is preserved while $\mathscr{F}_\pi$ and $\mathscr{C}_\pi$ are interchanged.

The next two lemmas recall some standard facts about the cohomology of groups. The groups $H^i(\pi, M)$ for $i < 0$ are defined as in [CE, Ch. XII] where full details of the proofs may be found.

Lemma 8.1. *Let π' be a subgroup of π.*

(a) *If M is a $\mathbf{Z}\pi$-module* $\operatorname{Ext}^i_{\mathbf{Z}\pi}(\mathbf{Z}\pi/\pi', M) \approx H^i(\pi', M)$ *for $i \geq 0$.*

(b) (Shapiro's Lemma). *If N is a $\mathbf{Z}\pi'$-module then*

$$H^i(\pi', N) = H^i(\pi, \operatorname{Ind}^\pi_{\pi'}(N)) \quad \textit{for all } i.$$

Proof. In (a) both sides are cohomological δ-functors and they agree for $i = 0$ ([G]). The same argument works for (b) if $i \geq 0$. To get the result for all i it is enough to check it for $i = 0$ and use the sequences $0 \to N \to \mathrm{Hom}(\mathbf{Z}\pi', N) \to X \to 0$ and $0 \to Y \to \mathbf{Z}\pi' \otimes N \to N \to 0$ to shift dimensions. Both functors vanish on the middle terms. □

Lemma 8.2 (Duality). (a) *If M is any $\mathbf{Z}\pi$-module,*

$$H^{p-1}(\pi, \mathrm{Hom}(M, \mathbf{Q}/\mathbf{Z})) \approx \mathrm{Hom}(H^{-p}(\pi, M), \mathbf{Q}/\mathbf{Z}).$$

(b) *If M is torsion free and finitely generated over $\mathbf{Z}$ then*

$$H^p(\pi, M) \approx \mathrm{Hom}(H^{-p}(\pi, M^\vee), \mathbf{Q}/\mathbf{Z}).$$

Proof. For (a) we need only check the result for $p = 1$ and use the dimension shifting technique, see [CE, Ch. XII, Th. 6.4]. In [CE, Ch. XII, Th. 6.6], (b) is only stated for the case $M = \mathbf{Z}$ but a similar proof works in general. We hom M into $0 \to \mathbf{Z} \to \mathbf{Q} \to \mathbf{Q}/\mathbf{Z} \to 0$ getting $0 \to M^\vee \to V \to \mathrm{Hom}(M, \mathbf{Q}/\mathbf{Z}) \to 0$. Since V is a rational vector space it has trivial cohomology so $H^{p-1}(\pi, \mathrm{Hom}(M, \mathbf{Q}/\mathbf{Z})) \xrightarrow{\approx} H^p(\pi, M^\vee)$ and we can use (a). □

We now turn to the results from [CTS].

Lemma 8.3. *Let $M \in \mathscr{L}_\pi$. Then*

(a) $M \in \mathscr{F}_\pi \Leftrightarrow H^{-1}(\pi', M) = 0$ *for all* $\pi' \subset \pi$.
(b) $M \in \mathscr{C}_\pi \Leftrightarrow H^1(\pi', M) = 0$ *for all* $\pi' \subset \pi$.

Proof. By Lemma 8.2(b) it will suffice to prove (b). Since Ext is additive, $M \in \mathscr{C}\pi \Leftrightarrow \mathrm{Ext}^1_{\mathbf{Z}\pi}(P, C) = 0$ for all $P = \mathbf{Z}\pi/\pi'$ so the result follows from Lemma 8.1(a). □

Lemma 8.4. $\mathscr{P}_\pi \subset \mathscr{F}_\pi \cap \mathscr{C}_\pi$.

Proof. Since $\mathscr{F}_\pi$ and $\mathscr{C}_\pi$ are closed under direct sums and direct summands it is enough to check that modules of the form $P = \mathbf{Z}\pi/\pi'$ lie in them. For $\mathscr{C}_\pi$ this follows from Lemma 8.3(b) and Shapiro's lemma since $P \approx \mathrm{Ind}^{\pi}_{\pi'}(\mathbf{Z})$ and $H^1(\pi', \mathbf{Z}) = \mathrm{Hom}(\pi', \mathbf{Z}) = 0$. For $\mathscr{F}_\pi$ we use the involution $M \mapsto M^\vee$ to reduce to the case just done. □

Lemma 8.5. *If $M \in \mathscr{L}_\pi$, there are exact sequences:*

(a) $0 \to M \to P \to F \to 0$ *with* $F \in \mathscr{F}_\pi$ *and* $P \in \mathrm{Perm}_\pi$.
(b) $0 \to C \to Q \to M \to 0$ *with* $C \in \mathscr{C}_\pi$ *and* $Q \in \mathrm{Perm}_\pi$.

Proof. For (b) let $Q = \bigoplus_{\pi' \subset \pi} \mathbf{Z}[\pi/\pi'] \otimes M^{\pi'}$ with the obvious map $\sigma \otimes m \to \sigma m$ onto M. Let C be the kernel. The exact cohomology sequence gives $H^1(\pi', C) = 0$ for all $\pi' \subset \pi$. We then get (a) by applying (b) to $M^\vee$ and taking duals. □

Corollary 8.6.

$$\begin{aligned}\mathscr{P}_\pi &= \{M \in \mathscr{L}_\pi \mid \mathrm{Ext}^1_{\mathbf{Z}\pi}(F, M) = 0 \text{ for all } F \in \mathscr{F}_\pi\} \\ &= \{M \in \mathscr{L}_\pi \mid \mathrm{Ext}^1_{\mathbf{Z}\pi}(M, C) = 0 \text{ for all } C \in \mathscr{C}_\pi\}.\end{aligned}$$

Proof. Using (a) of the lemma we see that $P \approx M \oplus F$. Similarly $Q \approx C \oplus M$ using (b). □

If $F_1, F_2 \in \mathscr{F}_\pi$ we say that F_1 and F_2 are stably equivalent if there are $S_1, S_2 \in \mathrm{Perm}_\pi$ such that $F_1 \oplus S_1 \approx F_2 \oplus S_2$. Let F_π be the monoid of stable equivalence classes so defined. If $M \in \mathscr{L}_\pi$ choose a resolution as in Lemma 8.5(a) and define $\rho(M) \in F_\pi$ to be the class of F. We will see below that these definitions agree with those of Section 7.

Lemma 8.7. *$\rho(M)$ is well defined.*

Proof. Suppose $0 \to M \to P' \to F' \to 0$. Following one of the standard proofs of Schanuel's lemma we form the pushout diagram

$$\begin{array}{ccccccccc} & & 0 & & 0 & & & & \\ & & \downarrow & & \downarrow & & & & \\ 0 & \to & M & \to & P & \to & F & \to & 0 \\ & & \downarrow & & \downarrow & & \| & & \\ 0 & \to & P' & \to & X & \to & F & \to & 0. \\ & & \downarrow & & \downarrow & & & & \\ & & F' & = & F' & & & & \\ & & \downarrow & & \downarrow & & & & \\ & & 0 & & 0 & & & & \end{array}$$

The middle row and column split giving $P \oplus F' \approx P' \oplus F$. □

Lemma 8.8. *Let $M, N \in \mathscr{L}_\pi$. Then $\rho(M) = \rho(N)$ if and only if there are short exact sequences*

$$0 \to M \to E \to P \to 0,$$
$$0 \to N \to E \to Q \to 0,$$

with $P, Q \in \mathrm{Perm}_\pi$.

Proof. Suppose $\rho(M) = \rho(N)$. Let $0 \to M \to P \to F \to 0$ and $0 \to N \to Q \to F' \to 0$ as in Lemma 8.5(a). We can assume that $F = F'$ by replacing $P \to F$ by $P \oplus S \to F \oplus S$ and similarly for $Q \to F'$. Form the pullback diagram

$$\begin{array}{ccccccccc} & & & & 0 & & 0 & & \\ & & & & \downarrow & & \downarrow & & \\ & & & & N & = & N & & \\ & & & & \downarrow & & \downarrow & & \\ 0 & \to & M & \to & E & \to & Q & \to & 0 \\ & & \| & & \downarrow & & \downarrow & & \\ 0 & \to & M & \to & P & \to & F & \to & 0. \\ & & & & \downarrow & & \downarrow & & \\ & & & & 0 & & 0 & & \end{array}$$

This gives the two required sequences. For the converse choose $0 \to E \to S \to F \to 0$ as in Lemma 8.5(a). Using $M \hookrightarrow E \hookrightarrow S$ we get $0 \to M \to S \to S/M \to 0$ and $0 \to E/M \to S/M \to F \to 0$. The latter sequence splits by the definition of $\mathscr{F}_\pi$ since

$E/M = P$ so $S/M \approx P \oplus F$. Therefore $\rho(M) = [S/M] = [F] \in F_\pi$ and similarly for N. □

It follows that the present definitions of F_π and $\rho(M)$ agree with those of Section 7.

The monoid F_π is difficult to determine explicitly. We can define some more easily calculable invariants in the following way. Suppose $\theta: \mathbf{Z}\pi \to R$ is a ring homomorphism with R a Dedekind ring. If M is in $\mathscr{L}_\pi$ define $(R \otimes_{\mathbf{Z}\pi} M)_0$ to be $R \otimes_{\mathbf{Z}\pi} M$ modulo its torsion submodule. If $M = \mathbf{Z}\pi/\pi'$ then M is a quotient of $\mathbf{Z}\pi$ so $R \otimes_{\mathbf{Z}\pi} M$ is a quotient of R and hence is either R itself or torsion. It follows that $(R \otimes_{\mathbf{Z}\pi} M)_0$ is free for $M \in \operatorname{Perm} \pi$ so we can define a homomorphism $c_\theta: F_\pi \to C(R)$ by sending F to the ideal class of the module $(R \otimes_{\mathbf{Z}\pi} M)_0$.

This construction is most interesting when π is abelian. In that case, the integral closure A of $\mathbf{Z}\pi$ in $\mathbf{Q}\pi$ is a product of rings of integers of cyclotomic fields $A = \prod A_i$ and we get $F_\pi \to \prod C(A_i)$. In general this is not injective. Endo and Miyata have shown it to be an isomorphism if π is a cyclic p-group ([EM]).

9. The Original Example

Let G be a cyclic group of prime order p. Then $\mathbf{Q}(x_1, \ldots, x_p)^G = \mathbf{Q}(K/\mathbf{Q}, M)$ as above where $K = \mathbf{Q}(\zeta_p)$ and $\pi = \operatorname{Gal}(K/\mathbf{Q}) = \mathbf{Z}/(p-1)$. From the construction in Section 7 we get $0 \to M \to X \to \hat{G} \to 0$ where $X = \mathbf{Z} \oplus \mathbf{Z}\pi$ as a π-module. This is easily seen from the fact that π permutes the non-trivial characters of G freely. This shows that $M = \mathbf{Z} \oplus I$ where $0 \to I \to \mathbf{Z}\pi \to \mathbf{Z}/p\mathbf{Z} \to 0$ with π acting on $\mathbf{Z}/p\mathbf{Z}$ by $\sigma(x) = rx$, r being a generator of $(\mathbf{Z}/p\mathbf{Z})^*$. It follows that $I = (\sigma - r, p)$. Since p is prime to the order of π, I is projective. To compute $\rho(M)$ we use $0 \to M \to \mathbf{Z} \oplus L \to P \to 0$ where $I \oplus P \approx L$ is free. Use c_θ where $\theta: \mathbf{Z}\pi \to R = \mathbf{Z}[\zeta_{p-1}]$. We get $c_\theta \rho(M) = \operatorname{cl}(R \otimes_{\mathbf{Z}\pi} P) = -\operatorname{cl}(R \otimes_{\mathbf{Z}\pi} I) = -\operatorname{cl} \mathfrak{p}$ where $\mathfrak{p} = (\zeta - r, p) \subset R$. If K is rational then $\mathfrak{p}$ must be principal so $\mathfrak{p} = (a)$ and $N_{R/\mathbf{Z}}(a) = \pm p$. If $p = 47$, this is impossible since $\pm p$ is not even a norm from $\mathbf{Z}[\sqrt{-23}] \subset R$. Saltman remarks that there is a generic Galois extension for G by Theorem 5.5 so such extensions may exist even if Noether's problem has a negative solution.

The above method fails for $G = \mathbf{Z}/8\mathbf{Z}$ since here π is the 4-group and the integral closure of $\mathbf{Z}\pi$ is $\mathbf{Z}^4$. In this case we can use instead the following remark which is used in [L].

Lemma 9.1. *If $Q(K/k, M)$ is rational and $M \in \mathscr{C}_\pi$ then $M \in \mathscr{P}_\pi$.*

Proof. By Lemma 8.8 we have $0 \to M \to E \to P \to 0$ with P and $E = Q$ in Perm_π. Since $M \in \mathscr{C}_\pi$ the sequence splits. □

10. Stable Equivalence

We say that two field extensions F_1 and F_2 of k are stably birationally equivalent if there are indeterminates x_i and y_j such that $F_1(x_1, \ldots, x_n) \approx F_2(y_1, \ldots, y_m)$. The invariant ρ of Section 7 is a stable birational invariant since if $A \subset K$ is chosen as in Section 7 we can use $A[x_i, \ldots, x_n] \subset K(x_1, \ldots, x_n)$. The following is proved in [CTS] using ideas from [L].

Theorem 10.1. *$Q(k'/k, M)$ and $Q(k'/k, N)$ are stably birationally equivalent if and only if $\rho(M) = \rho(N)$ in F_π.*

This follows easily from

Lemma 10.2. *Let $0 \to M \to N \to S \to 0$ with $S \in \mathrm{Perm}_\pi$. Then*

$$Q(k'/k, N) \approx Q(k'/k, M \times S).$$

Proof. Let $E = Q(k'/k, M)$. In $Q(k'/k, N)^*$ consider the exact sequence $1 \to E^* \to E^*N \xrightarrow{j} S \to 1$ where $j(en)$ is the image of n under $N \to S$. This sequence splits since $\mathrm{Ext}^1_{\mathbf{Z}\pi}(S, E^*) = 0$. It is sufficient to check this for $S = \mathbf{Z}\pi/\pi'$. By Lemma 8.1(a) it then reduces to Hilbert's Theorem 90. Therefore, $E^*N \approx E^* \times S$ and the lemma follows easily.

Now if $\rho(M) = \rho(N)$ we have $0 \to M \to X \to S \to 0$ and $0 \to N \to X \to T \to 0$ by Lemma 8.8 so $Q(k'/k, M \times S) \approx Q(k'/k, X) \approx Q(k'/k, N \times T)$. Lemma 7.1 shows that $Q(k'/k, M \times S)$ is rational over $Q(k'/k, M)$ and the theorem follows. □

11. Lenstra's Theorem

Let G be an abelian group acting simply transitively on the indeterminates $x_1, \ldots, x_n$. Let $k_G = k(x_1, \ldots, x_n)^G$. Lenstra [L] has given a complete answer to Noether's question in this case.

Theorem 11.1 (Lenstra). *Let G be a finite abelian group. The following conditions are equivalent:*

(1) k_G is rational over k.
(2) k_G is stably rational over k.
(3) (i) *The invariants $c_\theta \rho(k_G)$ are trivial.*
  (ii) *If* char $k = 2$, *let r be the largest power of 2 dividing the exponent of G. Then $k(\zeta_r)/k$ is cyclic.*

Condition (3)(ii) is, of course, the same one that came up in connection with Grunwald's theorem and in Saltman's work (Theorem 5.5). In Lenstra's work it comes in by way of Lemma 9.1 and a cohomology calculation. Lenstra expresses condition (3)(i) in a very explicit form. For the precise statement and for the proof of the theorem the reader is referred to Lenstra's paper ([L]) and to Kervaire's Bourbaki talk on it ([Ke]). Note that if the characteristic of k divides the order of G the definition of the invariant ρ must be modified since Fischer's theorem cannot be applied. Lenstra uses a result of Gaschütz to reduce to the case where $|G|$ is prime to char k. The arguments discussed here can then be applied.

12. Algebraic Tori

Voskresenskiĭ observed that the fields $Q(k'/k, M)$ considered above are exactly the function fields of algebraic tori. He discussed the relation between geometrically defined properties of tori and the algebraic material presented here and used this to obtain results on Noether's problem. Further work in this direction is due to Endo and Miyata. The papers of Endo and Miyata are particularly noteworthy for their detailed results on the structure of the monoid F_π. Some beautiful results in this direction have been obtained by Colliot-Thélène and Sansuc who use the ideas discussed here to study the relation of R-equivalence (a refinement of rational equivalence) on tori. The rest of this paper will be devoted to an exposition of some of their results.

If k is an algebraically closed field, a torus over k is an algebraic group isomorphic to a finite product of copies of the multiplicative group $\mathbf{G}_m(k) = k^*$. This can be embedded as an affine variety in affine $2n$-space with coordinate ring

$$k[x_1, x_1^{-1}, \ldots, x_n, x_n^{-1}].$$

If k is any field, a torus over k means a k-form of $(\mathbf{G}_m)^n$, i.e. an affine algebraic group T over k such that T becomes isomorphic to $(\mathbf{G}_m)^n$ over the algebraic closure $\bar{k}$ of k, $\bar{k} \otimes_k T \approx (\mathbf{G}_m)^n$. It is known that every torus splits over a finite separable extension K/k, i.e. $K \otimes_k T \approx (\mathbf{G}_m)^n$ ([Bo, Ch. III, Prop. 8.11]). For the present purposes we can just take this as part of the definition. We can also clearly assume that K/k is a Galois field extension. Let $T = \operatorname{Spec} A$ with the multiplication given by

$$\Delta: A \to A \otimes_k A$$

which induces $T \times T \to T$. Then $K \otimes_k A = K[x_1, x_1^{-1}, \ldots, x_n, x_n^{-1}] = K[M]$ where M is the multiplicative group generated by $x_1, \ldots, x_n$. It is easy to see that $M = \{u \in (K \otimes_k A)^* \mid \Delta u = u \otimes u\}$. In fact, M is the character group of $K \otimes T$ since $\operatorname{Hom}_{\mathrm{Alg\,gp}}(K \otimes T, \mathbf{G}_m) = \operatorname{Hom}_{\mathrm{bialg}}(K[x, x^{-1}], K \otimes A) = M$. Let π be the Galois group of K/k. Then $A = (K \otimes A)^\pi = K[M]^\pi = A(K/k, M)$ in the notation of Section 7. Therefore the k-tori are exactly the groups

$$T(K/k, M) = \operatorname{Spec} A(K/k, M)$$

for $M \in \mathscr{L}_\pi$, and the quotient fields of k-tori are exactly the fields $Q(K/k, M)$. If $\mathbf{G}$ is the Galois group of k_s/k where k_s is the separable closure of k, this gives us an equivalence between the category of k-tori and the category $(\mathscr{L}_G)^{\mathrm{op}}$ where $\mathscr{L}_G$ is the category of G-modules which are free and finitely generated as abelian groups and on which G acts continuously, i.e. the stabilizer of each element is an open subgroup ([Bo, Ch. III, §8]).

A group scheme T over k can be viewed as a functor from k-algebras to groups by setting $T(A) = \operatorname{Hom}_{\mathrm{Sch}/k}(\operatorname{Spec} A, T)$. If T is a torus, this can be given very explicitly as follows:

$$\begin{aligned} T(A) &= \operatorname{Hom}_{k\text{-alg}}(K[M]^\pi, A) = \operatorname{Hom}_{k\text{-alg}}(K[M]^\pi, K \otimes_k A)^\pi \\ &= \operatorname{Hom}_{K\text{-alg}}(K \otimes_k K[M]^\pi, K \otimes_k A)^\pi = \operatorname{Hom}_{K\text{-alg}}(K[M], K \otimes_k A)^\pi \\ &= \operatorname{Hom}_{Z}(M, (K \otimes_k A)^*)^\pi = \operatorname{Hom}_{\mathbf{Z}\pi}(M, (K \otimes_k A)^*). \end{aligned}$$

where the fourth equality uses Proposition 2.5.

One possible measure of the extent to which a variety X fails to be rational is given by the set $X(k)/\mathrm{Rat}$ of rational points of X modulo rational equivalence. Colliot-Thélène and Sansuc, following Manin, consider a finer equivalence relation, that of R-equivalence. We say that $a, b \in X(k)$ are directly R-equivalent if there is a rational k-variety U and a morphism $f: U \to X$ defined over k such that $a, b \in f(U(k))$. The equivalence relation generated by this is called R-equivalence and is clearly finer than rational equivalence. In this definition we could restrict U to open sets in affine space $\mathbf{A}^n$ since a rational variety is covered by such sets. By drawing lines in $\mathbf{A}^n$ we see that it would suffice to take U to be open in $\mathbf{A}^1$.

Now let T be a torus over k and let $R \subset T(k)$ be the set of points which are R-equivalent to 1. It is very easy to see that R is a subgroup of $T(k)$ and that the R-equivalence classes are exactly the cosets of R in $T(k)$.

Theorem 12.1 (Colliot-Thélène and Sansuc). *If T is any k-torus, $T(k)/R = T(k)/\mathrm{Rat}$. This group is completely determined by $\rho(T) \in F_\pi$.*

Here, of course, $\rho(T)$ means $\rho(M)$ where $T = T(K/k, M)$ and $\pi = \mathrm{Gal}(K/k)$. It follows from this and Theorem 10.1 that $T(k)/R$ is a stable birational invariant of T.

The group $T(k)/R$ is given explicitly as a cohomology group determined by $\rho(M)$. This can be expressed in a manner independent of the choice of the splitting field K as follows. If S is a torus defined over k let $H^i(k, S) = \varinjlim H^i(\mathrm{Gal}(L/k), S(L))$, the limit being taken over all finite Galois field extensions L of k.

Lemma 12.2. *Let $T = T(K/k, M)$ with $\pi = \mathrm{Gal}(K/k)$. Then:*

(a) $H^0(k, T) = T(k) = \mathrm{Hom}_{\mathbf{Z}\pi}(M, K^*)$.
(b) $H^1(k, T) = H^1(\pi, T(K)) = H^1(\pi, \mathrm{Hom}(M, K^*)) = \mathrm{Ext}^1_{\mathbf{Z}\pi}(M, K^*)$.

Proof. Let $L \supset K \supset k$ with $G = \mathrm{Gal}(L/k)$ and $N = \mathrm{Gal}(L/K)$ so that $\pi = G/N$. Then $H^0(G, T(L)) = \mathrm{Hom}(M, L^*)^G = \mathrm{Hom}(M, L^{*N})^{G/N} = \mathrm{Hom}_G(M, K^*) = T(k)$. For any G-module W we have ([HS, Ch. VI, 8.2])

$$0 \to H^1(\pi, W^N) \to H^1(G, W) \to H^1(N, W)^\pi \to H^2(\pi, W^N) \to H^2(G, W).$$

Applying this to $W = \mathrm{Hom}(M, L^*)$ and observing that $H^1(N, W) = 0$ by Hilbert's Theorem 90 since $W = L^* \times \cdots \times L^*$ as an N-module, we get the required result. The final equality follows as in Lemma 8.1(a). □

Now let $T = T(K/k, M)$ and choose a resolution $0 \to M \to P \to F \to 0$ as in Lemma 8.5(a) with $P \in \mathrm{Perm}_\pi$ and $F \in \mathscr{F}_\pi$. The cohomology sequence of $0 \to \mathrm{Hom}(F, K^*) \to \mathrm{Hom}(P, K^*) \to \mathrm{Hom}(M, K^*) \to 0$ can be written as

$$1 \to S(k) \to E(k) \to T(k) \xrightarrow{\delta} H^1(k, S) \to H^1(k, E) \to \cdots$$

by Lemma 12.2 where $S = T(K/k, F)$ and $E = T(K/k, P)$. This sequence is clearly independent of the choice of the splitting field K. Now $H^1(k, E) = \mathrm{Ext}^1_{\mathbf{Z}\pi}(P, K^*) = 0$ by Lemma 8.1(a) and Hilbert's Theorem 90 since P is a direct sum of modules of the form $\mathbf{Z}\pi/\pi'$. If we write $R' = \mathrm{im}[E(k) \to T(k)]$ we get $\delta: T(k)/R' \to H^1(k, S)$.

Lemma 12.3. $R' = R$.

Proof. Clearly $R' \subset R$ since E is rational by Lemma 7.1. For the converse, suppose $U = \operatorname{Spec} A$ is open in A^n, $A = k[x_1, \ldots, x_n, s^{-1}]$ and $f: U \to T$ with $1, a \in f(U(k))$. It is enough to consider this case since R is a subgroup of $T(k)$. We must show that f factors through E or, equivalently, that $f \in T(A)$ lies in the image of $E(A)$. Let $\varepsilon: A \to k$ represent a point $e \in E(k)$ with $f(e) = 1$. Then $\varepsilon \otimes 1: K \otimes A \to K$ induces a ladder of Ext sequences

$$\begin{array}{ccccc} \operatorname{Hom}_{\mathbf{Z}\pi}(P, (K \otimes A)^*) & \to & \operatorname{Hom}_{\mathbf{Z}\pi}(M, (K \otimes A)^*) & \to & \operatorname{Ext}^1_{\mathbf{Z}\pi}(F, (K \otimes A)^*) \\ \downarrow & & \downarrow & & \downarrow \theta \\ \operatorname{Hom}_{\mathbf{Z}\pi}(P, K^*) & \to & \operatorname{Hom}_{\mathbf{Z}\pi}(M, K^*) & \to & \operatorname{Ext}^1_{\mathbf{Z}\pi}(F, K^*), \end{array}$$

or

$$\begin{array}{ccccc} E(A) & \to & T(A) & \to & \operatorname{Ext}^1_{\mathbf{Z}\pi}(F, (K \otimes A)^*) \\ \downarrow & & \downarrow & & \downarrow \theta \\ E(k) & \to & T(k) & \to & \operatorname{Ext}^1_{\mathbf{Z}\pi}(F, K^*). \end{array}$$

Now the argument just preceding Definition 7.2 applied to $K[x_1, \ldots, x_n]$ and s shows that we have an exact sequence $0 \to K^* \to (K \otimes A)^* \to Q \to 0$ where Q is the permutation module corresponding to the factorization of s over K. This sequence is split by $\varepsilon \otimes 1: (K \otimes A)^* \to K^*$. Since $F \in \mathscr{F}_\pi$, $\operatorname{Ext}^1_{\mathbf{Z}\pi}(F, Q) = 0$ and it follows that the map θ above is an isomorphism. Since $f \in T(A)$ maps to $1 \in T(k)$ by the choice of ε, an easy diagram chase shows that f is in the image of $E(A)$ as required. □

We have therefore proved the following result.

Theorem 12.4 (Colliot-Thélène and Sansuc). $\delta: T(k)/R \approx H^1(k, S)$.

This implies the second statement of Theorem 12.1 since $H^1(k, S) = \operatorname{Ext}^1_{\mathbf{Z}\pi}(F, K^*)$ by Lemma 12.2. If F is replaced by $F \oplus P$ where P is a permutation module, the value of this Ext group does not change since $\operatorname{Ext}^1_{\mathbf{Z}\pi}(P, K^*) = 0$ as observed just before Lemma 12.3. In [CTS] it is shown that the map $\delta: T(k) \to H^1(k, S)$ factors through a map $A_0(T) \to H^1(k, S)$ where A_0 is the Chow group of 0-cycles modulo rational equivalence. It follows that rationally equivalent points of $T(k)$ have the same image under δ and therefore, by Theorem 12.4, that rational equivalence is the same as R-equivalence of $T(k)$. The reader is referred to [CTS] for the details.

Theorem 12.5 (Colliott-Thélène and Sansuc). *Let T be a k-torus. If k is finitely generated over the prime field then $T(k)/R$ is finite.*

Proof. Let $T = T(K/k, M)$ and choose $0 \to M \to P \to F \to 0$ as above. Let $A \subset K$ be a π-stable subring which is finitely generated over $\mathbf{Z}$ and whose quotient field is K. We can assume that A is normal since its normalization will have the same properties ([N]). It follows from the Mordell–Weil theorem that the class group $\operatorname{Cl}(A)$ is finitely generated ([R][LDG]). Therefore, by replacing A by a suitable localization $A[s^{-1}]$ we can assume that $\operatorname{Cl}(A) = 0$, i.e. if I_i generate $\operatorname{Cl}(A)$ make $I_i A[s^{-1}] = A[s^{-1}]$. By [R] or [LDG], A^* is a finitely generated abelian group. The sequence $0 \to A^* \to K^* \to D \to 0$, where D is the divisor group of A, induces $\text{Ext}^1_{\mathbf{Z}\pi}(F, A^*) \to \text{Ext}^1_{\mathbf{Z}\pi}(F, K^*) \to \text{Ext}^1_{\mathbf{Z}\pi}(F, D) \to \cdots$ but the last group is 0 since $F \in \mathscr{F}_\pi$ and D is a permutation module. The fact that D is not finitely generated causes no trouble since $\text{Ext}(F, -)$ preserves filtrated direct limits, F being finitely generated. Therefore $\text{Ext}^1_{\mathbf{Z}\pi}(F, K^*) = H^1(k, S)$ is finitely generated since $\text{Ext}^1_{\mathbf{Z}\pi}(F, A^*)$ is. Since $T(k)/R = H^1(k, S) = H^1(\pi, S(K))$ is always annihilated by the order of π, the result follows. □

This proof also works if k is a local field with finite residue field ([CTS, R8]). Here A^* is replaced by an open subgroup of K^* with trivial cohomology.

The above theorems are only a very small sample of the results contained in [CTS] to which the interested reader is referred for further information.

References

[5] E. Noether. Rationale Funktionkörper. *Jahresber. Deut. Math.-Verein*, **22** (1913), 316–319.

[11] E. Noether. Gleichungen mit vorgeschriebener Gruppe. *Math. Ann*, **78** (1918), 221–229.

[42] E. Noether. Der Hauptgeschlechtssatz fur relativ-Galoissche Zahlkörper. *Math. Ann.*, **108** (1933), 411–419.

[A] E. Artin. *Algebraic Numbers and Algebraic Functions*. Notes by I. Adamson. Gordon and Breach: New York, 1967.

[AT] E. Artin and J. Tate. *Class Field Theory*. Benjamin: New York, 1967.

[ATI] E. Artin and J. Tate. A note on finite ring extensions. *J. Math. Soc. Japan*, **3** (1951), 74–77.

[B] H. Bass. *Algebraic K-Theory*. Benjamin: New York, 1968.

[Bo] A. Borel. *Linear Algebraic Groups*. Notes by H. Bass. Benjamin: New York, 1969.

[CE] H. Cartan and S. Eilenberg. *Homological Algebra*. Princeton Univ. Press: Princeton, 1956.

[CHR] S. Chase, D. Harrison, and A. Rosenberg. Galois theory and cohomology of commutative rings. *Mem. Amer. Math. Soc.*, **52** (1965), 1–19.

[CTS] J-L. Colliot-Thélène and J-J. Sansuc. La R-équivalence sur les tores. *Ann. Sci. E.N.S.*, **101** (1977), 175–230.

[DMI] F. De Meyer and E. Ingraham. *Separable Algebras over Commutative Rings*. Lecture Notes in Math., 181. Springer-Verlag: Berlin, Heidelberg, New York, 1971.

[EM1] S. Endo and T. Miyata. Invariants of finite abelian groups. *J. Math. Soc. Japan*, **25** (1973), 7–26.

[EM2] S. Endo and T. Miyata. Quasi-permutation modules over finite groups, I, II. *J. Math. Soc. Japan*, **25** (1973), 397–421 and **26** (1974), 689–713.

[EM3] S. Endo and T. Miyata, On a classification of the function fields of algebraic tori. *Nagoya Math. J.*, **56** (1974), 85–104.

[F] E. Fischer. Die Isomorphie der Invariantenkörper der endlichen Abel'schen Gruppen linearer Transformationen. *Gött. Nachr.*, (1915), 77–80.

[Fr] A. Fröhlich. Algebraic number theory. In *Emmy Noether: A Tribute to Her Life and Work*. Edited by J. W. Brewer and M. K. Smith. Marcel Dekker: New York, 1981

[G] A. Grothendieck. Sur quelques points d'algebre homologique. *Tôhoku Math. J.*, **9** (1957), 119–221.

[H1] M. Hajja. On the rationality of monomial automorphisms. *J. Algebra*, **73** (1981), 30–36.

[H2] M. Hajja. Rational invariants of meta-abelian groups of linear automorphisms. *J. Algebra*, to appear.

[HS] P. J. Hilton and U. Stammbach. *A Course in Homological Algebra*, Springer-Verlag: New York, Heidelberg, Berlin, 1971.

[Ka] I. Kaplansky. *Commutative Rings*. Allyn and Bacon: Boston, 1970.

[Ke] M. Kervaire. Fractions rationnelles invariantes. Sem. Bourbaki, Exp. 445 (1973/1974).

[Ki] C. Kimberling. Emmy Noether and her influence. In *Emmy Noether: A Tribute to Her Life and Work*. Edited by J. W. Brewer and M. K. Smith. Marcel Dekker: New York, 1981.

[Ku] W. Kuyk. On a theorem of Emmy Noether. *Proc. Ned. Akad. Wetensch.*, Ser. A, **67** (1964), 32–39.

[LDG] S. Lang. *Diophantine Geometry*. Interscience: New York, 1962.

[L] H. W. Lenstra, Jr. Rational functions invariant under a finite abelian group. *Invent. Math.*, **25** (1974), 299–325.

[L1] H. W. Lenstra, Jr. Rational functions invariant under a cyclic group. *Queen's papers in Pure and Appl. Math.*, **54** (1980), 91–99.

[M] K. Masuda. Application of the theory of the group of classes of projective modules to the existence problem of independent parameters of invariant. *J. Math. Soc. Japan*, **20** (1968), 223–232.

[N] M. Nagata. *Local Rings*. Interscience: New York, 1962.

[R] P. Roquette. Einheiten und Divisorenklassen in endlich erzeugbaren Körpern. *Jahresber. Deut. Math.-Verein*, **60** (1958), 1–27.

[Sa] D. Saltman. Generic Galois extensions and problems in field theory. *Adv. Math.*, **43** (1982), 250–283.

[Sp] A. Speiser. Zahltheoretische Sätze aus der Gruppentheorie. *Math. Z.*, **5** (1919), 1–6.

[Sw] R. G. Swan. Galois Theory. In *Emmy Noether: A Tribute to Her Life and Work*. Edited by J. W. Brewer and M. K. Smith. Marcel Dekker: New York, 1981.

[SwI] R. G. Swan. Invariant rational functions and a problem of Steenrod. *Invent. Math.*, **7** (1969), 148–158.

[V1] V. E. Voskresenskii. Birational properties of linear algebraic groups. *Izv. Akad. Nauk SSSR Ser. Mat.*, **34** (1970), 3–19 = *Math. USSR Izv.*, **4** (1970), 1–17.

[V2] V. E. Voskresenskii. On the question of the structure of the subfield of invariants of a cyclic group of automorphisms of the field $Q(x_1, \ldots, x_n)$. *Izv. Akad. Nauk SSSR Ser. Mat.*, **34** (1970), 366–375 = *Math. USSR Izv.*, **4** (1970), 371–380.

[V3] V. E. Voskresenskii. Rationality of certain algebraic tori. *Izv. Akad. Nauk SSSR Ser. Mat.*, **35** (1971), 1037–1046 = *Math. USSR Izv.*, **5** (1971), 1049–1056.

[V4] V. E. Voskresenskii. Fields of invariants of abelian groups. *Usp. Mat. Nauk*, **28** (1973), 77–102 = *Russ. Math. Surveys*, **28** (1973), 79–105.

[V5] V. E. Voskresenskii. Stable equivalence of algebraic tori. *Izv. Akad. Nauk SSSR Ser. Mat.*, **38** (1974), 3–10 = *Math. USSR Izv.*, **8** (1974), 1–7.

[V6] V. E. Voskresenskii. Some problems in the birational geometry of algebraic tori. *Proc. Int. Cong. Math.*, Vancouver, 1974, 343–347.

[W1] S. Wang. A counterexample to Grunwald's theorem. *Ann. Math.*, **49** (1948), 1008–1009.

[W2] S. Wang. On Grunwald's theorem. *Ann. Math.*, **51** (1950), 471–484.

Noether Normalization*

JUDITH D. SALLY†

A remarkable fact about Emmy Noether's work in commutative ring theory is that the foundation for the general theory of ideals which she laid in the 1920s remains today essentially as she laid it. It is an understatement to say it is well known that in an abstract commutative ring the ascending chain condition (acc) on ideals is equivalent to the existence of a finite basis for each ideal and that primary decomposition exists in an abstract commutative ring with acc on ideals. These ideas and many others found in Noether's two most famous papers in commutative ring theory: "Idealtheorie in Ringbereichen" and "Abstrakter Aufbau der Idealtheorie in algebraischen Zahl-und Funktionenkörpern" were revolutionary in the 1920s but are very familiar today.

Less familiar, perhaps, is Noether's paper "Der Endlichkeitssatz der Invarianten endlicher linearer Gruppen der Charakteristik p." This paper illustrates the power of Noether's abstract methods. For, with these tools, she was able to complete the solution, which she had begun 10 years earlier, to Hilbert's Fourteenth Problem for finite groups. Noether used what has become known as the Noether Normalization Lemma to make a ring into a finitely generated module over a "better" ring thereby exploiting not only finiteness conditions but also the good properties of the better ring.

Recall what the Normalization Lemma says.

If K is a field and T is a finitely generated K-algebra, then there exist elements $X_1, \ldots, X_n$ in T algebraically independent over K such that T is integral over $S = K[X_1, \ldots, X_n]$.

* This work was partially supported by the Mary Ingraham Bunting Institute of Radcliffe College and the National Science Foundation.

† Department of Mathematics, Northwestern University, College of Arts and Sciences, Lunt Hall, Evanston, IL 60201, U.S.A.

Here is a sketch of one of the modern proofs. Let $T = K[y_1, \ldots, y_m]$. We may as well assume that $y_1, \ldots, y_m$ are not algebraically independent over K, otherwise the proof is finished. Thus there is a non-trivial relation

$$(*) \qquad y_1^{a_1} \cdots y_m^{a_m} + u_b y_1^{b_1} \cdots y_m^{b_m} + \cdots + u_c y_1^{c_1} \cdots y_m^{c_m} = 0,$$

where the u's are in K and a_1, say, is non-zero. We change variables by setting $z_2 = y_2 - y_1^{\alpha_2}, \ldots, z_m = y_m - y_1^{\alpha_m}$ with $\alpha_2, \ldots, \alpha_m$ as yet unspecified natural numbers. Then $T = K[y_1, z_2, \ldots, z_m]$ and (*) becomes

$$(**) \qquad y_1^{a_1 + \alpha_2 a_2 + \cdots + \alpha_m a_m} + u_b y_1^{b_1 + \alpha_2 b_2 + \cdots + \alpha_m b_m} + \cdots + u_c y_1^{c_1 + \alpha_2 c_2 + \cdots + \alpha_m c_m} + h(y_1, z_2, \ldots, z_m) = 0,$$

where the degree of h in y_1 is less than the degree in y_1 of the displayed terms in (**). If we choose $\alpha_2, \ldots, \alpha_m$ so that the exponents $(1, \alpha_2, \ldots, \alpha_m) \cdot (a_1, \ldots, a_m), \ldots, (1, \alpha_2, \ldots, \alpha_m) \cdot (c_1, \ldots, c_m)$ are distinct, then (**) is an equation showing that y_1 is integral over $K[z_2, \ldots, z_m]$. By induction on m, $K[z_2, \ldots, z_m]$ is integral over a polynomial subring S. Consequently, the transitivity of integral dependence implies that T is integral over S.

Actually, in "Der Endlichkeitssatz der Invarianten endlicher linearer Gruppen der Charakteristik p" this lemma, for infinite ground fields, appears in the body of the proof of what Noether called the Finiteness Criterion. Noether refers the reader to a paper ([Hb]) of Hilbert where the graded version of the lemma is stated and proved. For, in the geometric set-up where T is the homogeneous coordinate ring of an n-dimensional subvariety of r-dimensional projective space $\mathbf{P}^r$, the Normalization Lemma says that there is a linear subspace L of dimension $r - n - 1$ such that $L \cap X = \varnothing$ and that the projection from $L: X \to \mathbf{P}^n$ is a finite map; facts no doubt used by geometers of the time.

Thus it may well be that the real innovation was Noether's use of the lemma. She used it as a tool to linearize the finitely generated K-algebra T over its polynomial subring S and thereby pass finiteness from S to T as an S-module. One might think, then, of Noether Normalization as the technique of making a ring into a finitely generated module over a "better" ring.

The setting for Noether's Finiteness Criterion is as follows. K is a field and R is a sub-K-algebra of a field $K(Y_1, \ldots, Y_s)$ of rational functions over K. Recall that R need not be a finitely generated K-algebra as is the case, for example, with

$$R = K[X, X^2Y, X^3Y^2, \ldots, X^{n+1}Y^n, \ldots] \subset K(X, Y),$$

whereas if R were a field it would be finitely generated as a field over K (for a modern proof, cf. [L]).

Noether's Finiteness Criterion. *R is a finitely generated K-algebra if and only if R is integral over a subring T which is a finitely generated K-algebra.*

For the proof that the condition is sufficient, Noether assumed that such a T exists and proved the Normalization Lemma for T to get T integral over a polynomial subring S. Then she exploited two facts. The first is that the ascending chain condition on ideals in S can be transferred to sub-S-modules of any K-algebra $\bar{S}$ that is itself a finitely generated S-module. The second is that S has many good properties. In particular, the integral closure of S in a finite extension of the quotient field of S is a finitely generated S-module. This is proved by Artin and van der Waerden in the paper [A–vdW] which immediately precedes Noether's in the Nachrichten. The remainder of Noether's proof is a little like a diagram chase.

$$\begin{array}{ccccccc}
K \subset S = K[X_1, \ldots, X_n] & \subset & T & \subset & R & & \\
 & & & & & \diagdown\!\!\!\cap & \\
\cap & & & & & & \bar{S} \\
 & & & & & \swarrow\!\cap & \\
K(X_1, \ldots, X_n) & \subset & & & L \subset K(Y_1, \ldots, Y_s). & &
\end{array}$$

L is the quotient field of R and $\bar{S}$ is the integral closure of S in L. L, as a subfield of a finitely generated field, is finitely generated and algebraic, so L is a finite algebraic extension of $K(X_1, \ldots, X_n)$. Then $R \subset \bar{S}$ and $\bar{S}$ is a finitely generated S-module. By acc on S-submodules of $\bar{S}$, R is a finitely generated S-module and, therefore a finitely generated K-algebra.

Noether applies this finiteness criterion to prove that, over any field K, the ring of invariants of a finite group G acting linearly on the polynomial ring $R = K[Z_1, \ldots, Z_s]$ is a finitely generated K-algebra, thus giving an affirmative answer to Hilbert's Fourteenth Problem for finite groups. In 1916, for the case char $K \nmid$ order G, she explicitly calculated a generating set for the invariants, namely the $\binom{|G| + s}{s}$ sums $\sigma_m = \sum_{g \in G} m^g$, where m ranges over the monomials in $Z_1, \ldots, Z_s$ of degree at most $|G|$, the order of G. In 1926 she resolved the remaining case by using the more abstract methods she had developed in the intervening years. She showed that the ring of invariants R^G satisfies the finiteness criterion by showing that it is integral over the subring generated by the σ_m.

Since Noether's time, normalization has evolved as a fundamental tool in commutative algebra. For example, it is only a short step to proofs of some classical results such as Hilbert's Nullstellensatz. We will illustrate some more modern uses where new results have been obtained by further exploiting the linearization.

Suppose that T is a finitely generated graded K-algebra. Via normalization, T is a finitely generated graded module over a polynomial subring $S = K[X_1, \ldots, X_n]$. This linearization can be further utilized because we have available all the tools of homological algebra. For, by Hilbert's Syzygy Theorem, T has projective dimension at most n over S. In particular, T is a free S-module precisely when T is a so-called Cohen–Macaulay ring. With this in mind, suppose once more that we have a finite group G acting linearly on a polynomial ring $R = K[Z_1, \ldots, Z_s]$. We know R^G, the ring of invariants, is a finitely generated module over a polynomial subring $S = K[X_1, \ldots, X_n]$. The natural question to ask is: "Is R^G free over S?", i.e. "Is R^G Cohen–Macaulay?" The answer is "Yes" if char $K = 0$ (Hochster and Eagon [H–E]) and "No" if char $K = p > 0$ (Bertin [B]). Further questions related to the structure of R^G for finite groups and infinite groups satisfying certain properties have been the source of some beautiful theorems relating algebraic geometry, commutative algebra, and combinatorics, cf., for example, work of Hochster–Roberts, Kempf, Reisner, Stanley, and Watanabe, among others.

Returning once more to the general normalization setting with the finitely generated K-algebra T integral over $S = K[X_1, \ldots, X_n]$, another natural question to ask is: "Is S a direct summand of T as S-modules?" The answer is "Yes." This is a special case of Hochster's

Direct Summand Conjecture. *Let S be a regular Noetherian ring and let $R \supset S$ be an S-algebra which is a finitely generated S-module. Then S is a direct summand of R.*

Hochster proved the conjecture when S contains a field K([H_1]). If char $K = 0$, the trace map provides a splitting. The proof for char $K = p > 0$ is a good introduction to the techniques which Hochster and Peskins–Szpiro have applied with such power in their work on the important open conjectures in the homological theory of modules over commutative rings (cf. [H_2], [PS]). Hochster's technique might be described as making a ring "better" by finding a good module over it.

The normalization process can be used to study the structure of more general algebras. One such example is the

Theorem of Generic Freeness. *Suppose that S is a Noetherian domain and that $T \supset S$ is a finitely generated S-algebra. Then there exists a non-zero element f in S such that T_f is a free S_f-algebra.*

Thus the theorem states that the map Spec $T \to$ Spec S is faithfully flat off $V(f)$, the closed set of primes containing f. Faithful flatness is, of course, the highly desirable property which implements passage of properties from S to T and vice-versa.

If T were a finitely generated S-module, we would have a map $\psi: S^r \to T \to 0$ which becomes an isomorphism when we tensor with K, the quotient field of S. Since ker ψ is a finitely generated S-module such that ker $\psi \otimes_S K = 0$, we can find a non-zero element f in S such that ker $\psi_f = 0$. To tackle the general case where T is a finitely generated S-algebra, we first employ primary decomposition to argue that we can assume that T is a domain. We apply the Normalization Lemma to $T \otimes_S K$ over K to get a non-zero element g in S and a polynomial subalgebra $S[X_1, \ldots, X_n]$ of T such that T_g is integral over $S_g[X_1, \ldots, X_n]$, where $n = \dim T \otimes_S K$. The case $n = 0$ is the previously mentioned case of a finitely generated module. Suppose $n > 0$ and let r be the rank of the torsion free $S_g[X_1, \ldots, X_n]$-module T_g. Thus there is an embedding

$$0 \to S_g[X_1, \ldots, X_n]^r \xrightarrow{\varphi} T_g,$$

which becomes an isomorphism when we tensor with $K(X_1, \ldots, X_n)$. It follows that dim coker $\varphi \otimes_S K < n$ and the proof can be completed by induction. □

Normalization can also be used indirectly to study local Noetherian rings $(R, \mathbf{m})$. For attached to any such is its associated graded ring $G = R/\mathbf{m} \oplus \mathbf{m}/\mathbf{m}^2 \oplus \mathbf{m}^2/\mathbf{m}^3 \oplus \cdots$ which *is* a finitely generated $R/\mathbf{m}$-algebra. Information about R can be obtained by studying G as a finitely generated module over a polynomial subring $S = R/\mathbf{m}[X_1, \ldots, X_n]$. Unfortunately, it is rare that G is free over S, i.e. that G is

Cohen–Macaulay. It is difficult to discover even when G is a torsion free S-module. Such questions lead directly to more questions about the behavior of Hilbert functions of local rings, cf., for example, [Sy].

For complete local Noetherian rings there is an

Analytic Normalization Theorem. *Let* $(R, \mathbf{m})$ *be an n-dimensional complete local Noetherian ring containing a field K. Let* $X_1, \ldots, X_n$ *be elements of R which generate an* **m**-*primary ideal. Then R is a finitely generated module over the power series subring* $K[[X_1, \ldots, X_n]]$.

This result is due, in various forms, to Rückert [R], Krull [K], and Cohen [C]. Notice how completeness gives freedom of choice of the "variables" $X_1, \ldots, X_n$.

Thus, these examples serve to illustrate that, with Noether's instruction on how to use it to apply abstract methods in a very concrete way, normalization has become a standard technique in commutative algebra.

Bibliography

E. Noether. Idealtheorie in Ringbereichen. *Math. Ann.*, **83** (1921), 24–66.

E. Noether. Abstrakter Aufbau der Idealtheorie in algebraischen Zahlund Funktionenkörpern. *Math. Ann.*, **96** (1927), 26–61.

E. Noether. Der Endlichkeitssatz der Invarianten endlicher linearer Gruppen der Charakteristik *p*. *Nachr. v. d. Ges. d. Wiss. zu Göttingen* (1926), 28–35.

References

[A-vdW] E. Artin and B. L. van der Waerden. Die Erhaltung der Kettensätze der Idealtheorie bei beliebigen endlichen Körpererweiterungen. *Nachr. v. d. Ges. d. Wiss. zu Göttingen* (1926), 23–27.

[B] M.-J. Bertin. Anneaux d'invarients d'anneaux de polynomes, en caractéristique *p*, *C. R. Acad. Sci. Paris*, Sér. A–B, **264** (1967), A653–A656.

[C] I. S. Cohen. On the structure and ideal theory of complete local rings. *Trans. Amer. Math. Soc.*, **59** (1946), 54–106.

[Hb] D. Hilbert, Über die vollen Invariantensysteme, *Math. Ann.*, **42** (1893), 313–373.

[H1] M. Hochster, Contracted ideals from integral extensions of regular rings. *Nagoya Math. J.*, **51** (1973), 25–43.

[H2] M. Hochster, *Topics in the Homological Theory of Modules over Commutative Rings*, CBMS Conf. Ser. in Math., 24 (1974).

[H-E] M. Hochster and J. A. Eagon, Cohen–Macaulay rings, invariant theory, and the generic perfection of determinantal loci, *Amer. J. Math.*, **93** (1971), 1020–1058.

[K] W. Krull, Zum Dimensionsbegriff der Idealtheorie, *Math. Z.*, **42** (1937), 745–766.

[L] S. Lang, *Introduction to Algebraic Geometry*, New York: Interscience, 1958.

[P-S] C. Peskins and L. Szpiro, *Dimension Projective Finie et Cohomologie Locale*, Inst. Haute Études Sci. Publ. Math., 42, Paris, 1973, 323–395.

[R] W. Rückert, Zum Eliminationsproblem der Potenzreihenideale, *Math. Ann.*, **107** (1933), 259–281.

[Sy] J. D. Sally, *Numbers of Generators of Ideals in Local Rings*, Lecture Notes in Pure and Appl. Math., 35, Marcel Dekker: New York, 1978.

Some Non-commutative Methods in Algebraic Number Theory

Olga Taussky*

Introduction

Some time between the years 1930–32 I heard Emmy cry out: " $1 - S = 2$ if $S = -1$." What she meant was, of course, that the symbolic power $1 - S$ implies squaring if S is the automorphism given by the inverse. Many times I heard her say, in many contexts: "Das muss hyperkomplex bewiesen werden," using the word hyperkomplex as an adverb. Both of these utterances were crucial for the work of Emmy that fits into the title of this article. Their implications illuminate a vast area of methods, formulations, new ideas.

One of Emmy's major hopes was to use her algebraic methods to reprove theorems of algebraic number theory and obtain a better insight for them in this way. One of the last theses produced under her guidance, the one by O. Schilling, does have a title related to this article. Her lecture at the International Congress of Zürich in 1932 shows her achievements. The phrase "Emmy was way ahead of her time" is used frequently in this area.

There are two research articles written about this in a modern set up, one by Martha Smith, published in a Newsletter of the Association of Women in Mathematics, November 1976, the other by A. Fröhlich's chapter on "Algebraic number theory," in *Emmy Noether, A Tribute to Her Life and Work*, Marcel Dekker, 1981, 157–163, giving details, partially complementing each other.

One of these achievements is the formulation and proof of the Hauptgeschlechtsatz, the principal genus theorem, an achievement that was very dear to her and admired by her colleagues. The concepts and methods used for this were very close to cohomology and can be looked upon as a major advance towards this theory. It is visible in the treatment. Only the words need to be replaced. But it

* Department of Mathematics, 253-37, CALTECH, Pasadena, CA 91125, U.S.A. The author has received advice for the presentation of this article from E. C. Dade, D. Estes, R. Guralnick, and H. Kisilevsky.

seems that the explicit replacement has only been in the two publications mentioned above.

Since I knew Emmy personally at this time of her life and since she turned the conversation several times to this topic—not necessarily in technical terms—discussing even the preparation of her Zürich lecture, I have chosen this topic as the major item of this article. However, I do not plan to give a description of the vast research that grew out of it or out of its reincarnations. For, there is a strange happening to be observed in the history of this. As I said earlier, this work was looked at as ahead of her time. But the time came for it, starting about 20 years later, to people using cohomology explicitly or implicitly. In two papers by Arnold Scholz (1936, 1940)—also ahead of his time—related methods are used when generalizing Hasse's results on norms in cyclic fields, a fact connected with principal genus theory. In recent years Jehne realized the value of this work, referred to in the bibliography of the author's obituary notice for Scholz. A great debt is owed to Jehne for his and his school's unearthing this, adding further development. A. Fröhlich's thesis in the late 1940s developed his own theory of principal genus independently. In the early 1950s a true blossoming broke out. Terada developed the principal genus for abelian fields in 1952 and in 1953 Kuniyoshi and Takahashi for the general case. In contrast to Emmy, they treat the subject inside the field and do not go to algebras connected with it.

Applications of cohomology in algebraic number theory broke through from then on. Still in the early 1950s there is work by Hochschild and Nakayama (1952). There is the work by Leopoldt (1953). Weil (1951) stressed its importance. Following the work of Artin and Tate cohomology became finally embedded in class field theory, in the 1950s. But non-abelian class field theory needed the work of Langlands and Tate, decades later.

Returning to the history of the principal genus itself another fact has to be noted. Different people, starting with Emmy, call their theorems the principal genus theorem. It seems that any definition, any result generalizing the relative cyclic case is acceptable. In the cyclic case the principal genus can be defined as the ideal class of $\mathscr{A}^{1-\sigma}$, σ an element of the cyclic Galois group and these ideals can be characterized by the fact that their norms are also norms of an element of the field. This brings me back to the first sentence of this article: In the case of quadratic fields the principal genus is given by the square classes. It was Gauss who started us on this and I will start by taking my orders from him. But I will return to Emmy by linking up with cyclic algebras, a favorite subject of hers and by linking them up with central polynomials, a modern concept.

Gauss characterized his principal genus, the square classes of quadratic forms, via characters, and Emmy studied relative normal extensions of degree n with Galois group G. She gave several characterizations of "principal genus" and, in particular, gave "cohomological" conditions on sets of n ideals which finally implied the existence of a single ideal $\mathscr{A}$ such that the ideals come from the classes of $\mathscr{A}^{1-\sigma}$, $\sigma \in G$. This fact is obtained in the language of inner automorphism.

Gauss did not have Galois theory, nor ideals in quadratic fields, nor matrices. But he started it all. But if you have the power of these concepts you obtain a better understanding.

This article has no connection with the theories mentioned here. It deals with the principal genus in normal (and finally even general) algebraic number fields, by methods of matrix theory, linear algebra and central polynomials. These methods had been employed by the author previously for the study of composition of binary quadratic and even n-ary forms of degree n. This will be repeated here briefly since it links with the methods for the principal genus. Two new characterizations for the latter are then obtained (Theorems A, B), the latter leading to a generalization of the principal genus. Theorem B is linked to Formanek's (1972) central polynomial. This theorem is then the main result as of now.

1. Composition of Integral Binary Quadratic Forms

Gauss, a smart man as he was, somehow turned the problem of squaring forms upside down.

He showed not only how to square a form, but also how to multiply them, by a process he called composition.

Let $f(x_1, \ldots, x_n)$, $g(y_1, \ldots, y_n)$ be two quadratic forms. Then under certain circumstances the biquadratic polynomial fg can again be expressed as a quadratic form $h(z_1, \ldots, z_n)$ where z_i are bilinear expressions $z_j = \sum c_{ikj} x_i y_k$. In particular, he studied pairs of binary integral quadratic forms

$$a_1 x^2 + b_1 xy + c_1 y^2, \qquad a_2 x^2 + b_2 xy + c_2 y^2$$

(actually he assumed the b_i even) with the same discriminants, assumed not a square, $b_i^2 - 4a_i c_i$.

I will rewrite them as

$$a_{21}x^2 + (a_{22} - a_{11})xy - a_{12}y^2, \qquad b_{21}x^2 + (b_{22} - b_{11})xy - b_{12}y^2.$$

Then I build up two matrices

$$A = \begin{pmatrix} a_{11} & a_{12} \\ a_{21} & a_{22} \end{pmatrix}, \qquad B = \begin{pmatrix} b_{11} & b_{12} \\ b_{21} & b_{22} \end{pmatrix},$$

and arrange it so that not only

$$(a_{11} - a_{22})^2 + 4a_{12}a_{21} = (b_{11} - b_{22})^2 + 4b_{12}b_{21},$$

but also

$$a_{11} + a_{22} = b_{11} + b_{22},$$

$$a_{11} - a_{22} = b_1, \qquad b_{11} - b_{22} = b_2.$$

In these circumstances it follows that A, B have identical characteristic polynomials and that A, B are similar over Q:

$$TAT^{-1} = B, \ T \text{ integral.}$$

This links Gauss' work to similarity of matrices. This will be studied in the next section. The most general T which comes in question here is given by a matrix whose entries are linear forms with integral coefficients in two parameters λ, μ. The determinant of such a matrix is an integral quadratic form. It can then be shown that this form is the composition of two quadratic forms which are related to the forms in question. For further details, see Taussky (1981).

2. Similarity of Integral Matrices

Let A, B be $n \times n$, integral matrices, with the same irreducible characteristic polynomial $f(x)$ (although free from multiple zeros would be sufficient). I assume α to be a zero of $f(x)$. I then study the integral matrices T for which

$$(*) \qquad TAT^{-1} = B.$$

I have two objects in this study: to link T to ideals in the field $Q(\alpha)$ and to investigate det T. This will turn out close to the principal genus.

For this purpose a theorem, pointed out by the author in 1949, with a slightly different set up, was introduced by her as the theorem of Latimer and MacDuffee (1933). It is not entirely novel, actually.

This theorem does not assume $Q(\alpha)$ to be normal, nor $Z[\alpha]$ to be the maximal order. It deals with the set of all $n \times n$ integral matrices A for which $f(A) = 0$. Any two of such matrices are similar and are said to lie in the same class if they are similar over Z, or in other words if they are similar via a unimodular matrix. The number of these classes is finite. (There is work by Schur (1922) and Zassenhaus (1938) there).

Theorem 1 (Latimer and MacDuffee (1933)). *There is an explicit* 1–1 *correspondence between the matrix classes and the ideal classes given via*

$$(1) \qquad A\begin{pmatrix}\alpha_1\\ \vdots\\ \alpha_n\end{pmatrix} = \alpha\begin{pmatrix}\alpha_1\\ \vdots\\ \alpha_n\end{pmatrix},$$

where the α_i *form a* Z*-basis for an ideal in* $Z[\alpha]$.[1]

The theorem can also be presented in the form

$$p[A]\begin{pmatrix}\alpha_1\\ \vdots\\ \alpha_n\end{pmatrix} = p[\alpha]\begin{pmatrix}\alpha_1\\ \vdots\\ \alpha_n\end{pmatrix},$$

$p(x)$ any integral polynomial.

The following theorem concerns a pair A, B with $f(A) = f(B) = 0$ with corresponding ideals $\mathscr{A}$ with Z-basis $\alpha_1, \ldots, \alpha_n$, and $\mathscr{B}$ with Z-basis $\beta_1, \ldots, \beta_n$.

[1] Generalizations of this theorem to abstract situations have been obtained by Estes and Guralnick (1982).

Theorem 2 (Taussky (1981)). *There is a 1–1 correspondence between the solutions* T_ρ *of* (*) *and the elements* ρ *of the ideal* $\mathscr{A}\mathscr{B}^{-1}$ *(provided that* $\mathscr{B}^{-1}$ *exists) such that*

$$\rho\begin{pmatrix}\beta_1\\ \vdots\\ \beta_n\end{pmatrix} = T_\rho\begin{pmatrix}\alpha_1\\ \vdots\\ \alpha_n\end{pmatrix}. \tag{2}$$

This can be considered as a generalized eigenvalue–eigenvector situation. Observe that (1) is a special case of (2) in case $A = B$. For this implies $\mathscr{A} = \mathscr{B}$, hence T_ρ is the set of all integral polynomials $p(x)$ and $\rho = \{p(\alpha)\}$. However, (1) also implies (2). This is shown in Taussky (1966) for a special case and in Taussky (1981) transferred to the general situation.

3. Exploiting the Transposed Matrix

Theorem 3 (Taussky (1957)). *Let* $\mathscr{A}$ *be an ideal such that* $\mathscr{A}^{-1}$ *exists, then*

$$\begin{aligned}&\textit{ideal class } \mathscr{A} \rightleftarrows \textit{matrix class } A\\ \Rightarrow\ &\textit{ideal class } \mathscr{A}^{-1} \rightleftarrows \textit{matrix class } A'.\end{aligned}$$

Theorem 4 (Taussky (1966)). *The study of the equation*

$$TAT^{-1} = A' \tag{3}$$

implies that the set of elements ρ *in* (2) *is* $\mathscr{A}^2$.

This is a principal genus statement for $n = 2$.

Theorem 5 (Taussky (1962)). *The matrix* T *in* (3) *satisfies*

$$\det T = (-1)^{n(n-1)/2} \text{ norm } \lambda \cdot a^2, \tag{4}$$

where $\lambda \in Q(\alpha)$ *and* $a \in Z$.

This was reproved in slightly generalized form by Bender (1974) and Weiss (1982).

For $n = 2$ we have

$$\det T = -\text{norm } \lambda.$$

This result for $n = 2$ is an easy example of a fact known for cyclic fields namely, the principal genus consists of the ideals whose norm is equal to the norm of an element (in the case $n = 2$ the ideals $\mathscr{A}^2$). This is a consequence of other characterizations of the principal genus in the cyclic case which lead to norm residues which in this case are known to imply norms. The concept of cyclic fields brings me to the next section.

4. Normal Fields Over Q

For $n = 2$ the inverse ideal can be obtained also by conjugation under the Galois group. This is not always true for $n > 2$. However, the transposed matrix exists for all n's.

Let G be the Galois group of $Q(\alpha)$ and $\{\sigma\}$ the elements of G. The zeros of $f(x)$ are then $\{\alpha^\sigma\}$. Denote by $p_\sigma(x)$ the polynomial of degree $< n$ for which $p_\sigma(\alpha) = \alpha^\sigma$. It is called a Galois polynomial or root polynomial. Under the assumption of $Z[\alpha]$ to be the maximal order its coefficients are integers. Let $f(A) = 0$, A an $n \times n$ integral matrix. Because of the isomorphism of $Z[\alpha]$ and $Z[A]$, $p_\sigma(A)$ is again an integral matrix root of $f(x) = 0$. Observe that $p_\sigma(A)$ commutes with A so that normality can be replaced here by commutativity.

For the converse holds as well. One way of showing this is to use the fact that a matrix B which commutes with a matrix A with distinct eigenvalues is simultaneously diagonable with A and also that it is a polynomial in A.

It is easy to see that $p_\sigma(p_\tau)(A) = p_{\sigma\tau}(A)$.[1]

We can now ask: What matrix class corresponds to the ideal $\mathscr{A}^\sigma$ under the Latimer and MacDuffee theorem?

Theorem 6.[2] *The following correspondence holds:*

$$\begin{aligned}&\text{ideal class } \mathscr{A} \rightleftarrows \text{matrix class } A\\ \Rightarrow\ &\text{ideal class } \mathscr{A}^\sigma \rightleftarrows \text{matrix class } p_{\sigma^{-1}}(A).\end{aligned}$$

Proof. Assuming

$$A\begin{pmatrix}\alpha_1\\ \vdots\\ \alpha_n\end{pmatrix} = \alpha\begin{pmatrix}\alpha_1\\ \vdots\\ \alpha_n\end{pmatrix},$$

we have

$$p(A)\begin{pmatrix}\alpha_1\\ \vdots\\ \alpha_n\end{pmatrix} = p(\alpha)\begin{pmatrix}\alpha_1\\ \vdots\\ \alpha_n\end{pmatrix}$$

for any integral polynomial $p(x)$.

Choose $p = p_\tau$, the Galois polynomial:

$$p_\tau(A)\begin{pmatrix}\alpha_1\\ \vdots\\ \alpha_n\end{pmatrix} = p_\tau(\alpha)\begin{pmatrix}\alpha_1\\ \vdots\\ \alpha_n\end{pmatrix}.$$

Now apply σ to this equation:

$$p_\tau(A)\begin{pmatrix}\alpha_1^\sigma\\ \vdots\\ \alpha_n^\sigma\end{pmatrix} = p_\tau(\alpha^\sigma)\begin{pmatrix}\alpha_1^\sigma\\ \vdots\\ \alpha_n^\sigma\end{pmatrix}.$$

[1] Replacing $p_\sigma(A)$ by $X_\sigma^{-1}AX_\sigma$ for a suitable matrix X_σ it is known that the X_σ form a factor set, hence are linked up with the cohomology group H^2.

[2] This was also observed by P. Morton.

setting τ equal to σ^{-1} we obtain

$$p_{\sigma^{-1}}(A)\begin{pmatrix}\alpha_1^\sigma\\ \vdots\\ \alpha_n^\sigma\end{pmatrix} = p_{\sigma^{-1}}(\alpha^\sigma)\begin{pmatrix}\alpha_1^\sigma\\ \vdots\\ \alpha_n^\sigma\end{pmatrix} = \alpha\begin{pmatrix}\alpha_1^\sigma\\ \vdots\\ \alpha_n^\sigma\end{pmatrix}.$$

and we are back to the Latimer and MacDuffee situation. Hence $\mathscr{A}^\sigma$ corresponds to $p_{\sigma^{-1}}(A)$. This will now be employed to study the principal genus. □

5. The Principal Genus Expressed Via Similarity of Matrices

Theorem A. *The solutions T of the matrix equation*

$$TAT^{-1} = p_{\sigma^{-1}}(A)$$

are in 1–1 correspondence with the principal genus $\mathscr{A}(\mathscr{A}^\sigma)^{-1}$.

This follows immediately from Theorems 1 and 6.

Remark 1. Although the ideal $\mathscr{A}^{1-\sigma}$ corresponds to the composition of the matrix classes of A and $p_\sigma(A')$ it makes sense to investigate the matrix product $A(p_\sigma(A))^{-1}$, in spite of the fact that this is not connected with the research reported here. It has a Theorem 90 flavor. In the case of real or complex entries the matrices $A(A')^{-1}$ or $A(A^*)^{-1}$ have been studied by several people. They are called cosquares. Every unitary matrix U can be expressed as $A(A^*)^{-1}$ while it is also considered as a matrix of "norm" equal to I, norm meaning UU^*. Here A, A^* commute. Similarly, the ideals $\mathscr{A}^{1-\sigma}$ have norm $= 1$ and the Hilbert Theorem 90 holds there too. The matrices $A(p_\sigma(A))^{-1}$ have det equal to 1, A and $p_\sigma(A)$ commute, their eigenvalues are $\alpha_i(p_\sigma(\alpha_i))^{-1}$. So there are analogues.

Remark 2. The equation

$$AX - Xp(A) = 0$$

can be studied for all matrices A and polynomials $p(x)$. This was done to some degree in the Caltech thesis of J. A. Parker, 1976.

But it can also be considered as a generalization of Galois theory. For, let A be a rational matrix with irreducible characteristic polynomial $f(x)$. Then X non-singular is equivalent with the fact that $p(x)$ is a Galois polynomial for $f(x)$. If $Q(A)$ is a cyclic field then $\pm$ det X is a norm from the corresponding number field. For if S is a matrix which diagonalizes A then $S^{-1}XS$ is monomial (this fact is greatly generalized in the above-mentioned thesis) and the non-zero elements in the monomial are conjugate.

Cyclic fields can be linked profitably to cyclic algebras, a preferred concept of Emmy's. In fact Estes reproved the above observation by this method.

6. Links of det$(AB - BA)$ for 2×2 Rational Matrices with Gauss Principal Genus, Cyclic Algebras and Central Polynomials

For several years the author has observed facts concerning pairs of 2×2 rational non-commuting matrices which led to squares of ideals.

Estes (1979) has written an article interlinking them in a more abstract form. Only one of them is picked out here.

Theorem 7 (Taussky (1975)). *Let A, B be a pair of 2×2 rational matrices, and one of them, say A, with irrational eigenvalues α, α'. Then*

$$-\det(AB - BA) = \text{norm } \lambda, \qquad \lambda \in Q(\alpha),$$

and every norm from $Q(\alpha)$ can be obtained in this way.

The following facts then hold:

(i) The above relation leads to the intersection of the norms from two quadratic fields. For, in case B satisfies the same conditions as A and has eigenvalues β, β', then $-\det(AB - BA) = \text{norm } \mu$, $\mu \in Q(\beta)$.

(ii) Let A, B be integral and study the set of λ's obtained in this way for A fixed, but B variable. Then the λ's can be so chosen that they run through the square of the ideal $\bar{\mathscr{A}}$ where $\mathscr{A}$ is an ideal associated with the matrix A and $\bar{\mathscr{A}}$ is its conjugate.

Among the several proofs given for the initial theorem is one via cyclic algebras. This proof was initiated by Zassenhaus (1977), but rearranged by others. This proof led to a generalization for $n > 2$ (Taussky 1978). For, use the isomorphism of $Q(\alpha)$ and $Q(A)$ and the fact that $AB - BA$ is non-singular (Taussky 1982). Then $AB - BA$ plays the role of the automorphism of the field $Q(A)$ if A is replaced by $A - \operatorname{tr} A \cdot I$ which is no restriction. Further $(AB - BA)^2$ lies in $Q(A)$, for it is $-\det(AB - BA)I$. Hence $-\det(AB - BA)$ is a norm from $Q(\alpha)$, by classical facts that hold for cyclic algebras. This proof is due to Taussky and Kisilevsky, see Taussky (1982). What is displayed here is the fact that $(AB - BA)^2$ is a 2×2 central polynomial, i.e. a polynomial in two matrix variables which becomes a scalar matrix, namely, $-\det(AB - BA)I$ for every choice of the pair. (Kaplansky (1970) encouraged the study of this for $n > 2$. It was shown by Formanek (1972) and Razmyslov (1973) that such polynomials exist for all n's.)

7. Generalizations to Polynomials for Pairs of Matrices in the Case $n > 2$

The author studied in particular Formanek's (1972) polynomial, with a result for $n = 3$.

In particular property (ii) mentioned in connection with $-\det(AB - BA)$ can be generalized to Formanek's polynomial for two $n \times n$ matrices. As in the 2×2

case there is an ideal associated with the matrix A, derived from the $(1, i)$ entry in the appropriate matrix. This will now be described in detail. As in the 2×2 case it is attached to the principal genus and leads to a new characterization and generalization for it.

Theorem B. *Let A be an integral $n \times n$ matrix with eigenvalues $\{\alpha^{(i)}\}$ where $\alpha = \alpha^{(1)}, \alpha^{(2)}, \ldots, \alpha^{(n)}$ are the eigenvalues of A, which is assumed to have an irreducible characteristic polynomial. Let $\{B\}$ be the set of all integral $n \times n$ matrices. Let $\{\sigma\}$ be the family G of all embeddings of $Q(\alpha)$ in the complex numbers. Let S be a matrix $(S_{i\sigma})$ with $S_{i\sigma} = \alpha_i^\sigma$ which transforms A into diagonal form $\begin{pmatrix} \ddots & & \\ & \alpha^\sigma & \\ & & \ddots \end{pmatrix}$. Normalize S by assuming that its columns are the conjugates of a Z-basis for an ideal $\mathscr{A}$ corresponding to A in the sense of (1). Let $\beta_1, \ldots, \beta_n$ be the dual (fractional) ideal $\mathscr{A}'$ defined by trace $(\alpha_i \beta_k) = \delta_{ik}$. Then $S^{-1} = (\beta_j^\tau)$. Here σ, τ run over G and i, j run over $\{1, \ldots, n\}$. Let $\{\tilde{B}\}$ run through all $n \times n$ matrices $S^{-1}BS$. Then the set of all $\tilde{b}_{1j}$ is the set of ideals $\{\mathscr{A}'\mathscr{A}^\sigma\}$, $\sigma \in G$.*

Proof. By matrix multiplication it follows that $\tilde{b}_{1j} = \sum \beta_r^{(1)} b_{rs} \alpha_s^{(j)}$ where the $\beta_r^{(1)}$ refer to the first row of S^{-1} and the $\alpha_s^{(j)}$ to the jth column of S. Then use the fact that the b_{rs} run independently through all of Z. Hence $\tilde{b}_{1i}$ is the product of the two ideals mentioned. □

Remark 3. If $\mathscr{A}'$ is the inverse of $\mathscr{A}$ then $\tilde{b}_{1i}$ run through the ideals $\mathscr{A}^{-1}\mathscr{A}^\sigma$ which for a normal field are the inverses of the elements of the principal genus.

Remark 4. Theorem B explains why in (ii) the ideal $\bar{\mathscr{A}}$ turns up instead of $\mathscr{A}$.

Remark 5. Let B be another $n \times n$ integral matrix with irreducible characteristic polynomial then Theorem B leads to an interaction between the fields generated by the zeros of their characteristic polynomials.

Remark 6. In case of a non-normal field the product $\mathscr{A}'\mathscr{A}^\sigma$ has to be interpreted as an example of a product of two ideals, $\mathscr{A}$ with basis $\alpha_1, \ldots, \alpha_n$; $\mathscr{B}$ with basis $\beta_1, \ldots, \beta_n$ and product $\sum a_{ik} \alpha_i \beta_k$, $a_{ik} \in Z$.

References

E. Artin and J. Tate. *Class Field Theory*. Harvard Math. Dept., 1961.

E. A. Bender. Characteristic polynomials of symmetric matrices, II. *Linear and Multilinear Alg.*, **2** (1974), 55–63.

S. Eilenberg and S. Maclane. Cohomology theory in abstract groups. *Ann. Math.*, **48** (1947), 51–78.

S. Eilenberg and S. Maclane. Cohomology and Galois Theory, 1. *Trans. Amer. Math. Soc.*, **64** (1948), 1–20.

D. Estes. Determinants of Galois automorphisms of maximal commutative rings of 2 × 2 matrices. *Linear Alg. and Appl.*, **27** (1979), 225–243.

D. Estes. Oral communication, 1982.

D. Estes and R. Guralnick. Representations under ring extensions: Latimer–MacDuffee and Taussky correspondences.

D. K. Faddeev. On the representation of algebraic numbers by matrices. *Zap. Nauk. Sem. Leningrad Otd. Math. Inst.*, **46** (1974), 89–91.

P. Flor. Über die Festlegung eines Galoispolynoms durch seine Wurzelpolynome. *Monatsh. Math.*, **73** (1969), 397–311.

E. Formanek. Central polynomials for matrix rings. *J. Algebra*, **23** (1972), 129–132.

A. Fröhlich. The genus field and genus group in finite number fields, I. *Mathematika*, **6** (1959), 40–46; II. *ibid*, **6** (1959), 142–146.

Y. Furuta. The genus field and genus number in algebraic number fields. *Nagoya Math. J.*, **29** (1967), 281–285.

C. Gauss. *Disquisitiones Arithmeticae.*

K. Girstmair. On root polynomials of cyclic cubic equations. *Arch. Math.*, **36** (1981a), 313–326.

K. Girstmair. On relations between zeros of Galois polynomials. *Monatsh. Math.*, **91** (1981b), 203–214.

S. Gurak. On the Hasse norm principle. *J. reine angew. Math.*, **299–300** (1978), 10–27.

H. Hasse. Beweis eines Satzes und Widerlegung einer Vermutung über das allgemeine Normenrest—symbol. *Werke*, **I**. (1969), 155–160.

H. Hasse. A supplement to Leopoldt's theory of genera in abelian number fields. *J. Number Theory*, **1** (1969), 4–7.

F. P. Heider. Strahlknoten und Geschlechterkörper mod *m*. *J. reine angew. Math.*, **320** (1980), 52–67.

G. Hochschild and T. Nakayama. Cohomology in class field theory. *Ann. Math.*, **55** (1952), 348–366.

M. Ishida. *The Genus Fields of Algebraic Number Fields.* Springer Lecture Notes in Math, 555. Springer-Verlag: New York, 1976.

W. Jehne. Idealklassenfaktorensysteme und verallgemeinerte Theorie der verschränkten Produkte. *Math. Sem. Hamb.*, **18** (1952), 70–98.

W. Jehne. On knots in algebraic number theory. *J. reine angew. Math.*, **311** (1979), 215–254 (with many references).

W. Jehne. Der Hassesche Normensatz und seine Entwicklung. Preprint.

I. Kaplansky. Problems in the theory of rings. *Amer. Math. Monthly*, **77** (1970), 445–454.

H. Kleiman. The determination of a Galois polynomial by its root polynomials. *Amer. Math. Monthly*, **74** (1968), 55–56.

H. Kleiman. Methods for uniquely determining Galois polynomials and related theorems. *Monatsh. Math.*, **73** (1969), 63–68.

H. Kuniyoshi and S. Takahashi. On the principal genus theorem. *Tohoku Math. J.*, **5** (1953), 129–131.

C. G. Latimer and C. C. MacDuffee. A correspondence between classes of ideals and classes of matrices. *Ann. Math.*, **34** (1933), 313–316.

H. W. Leopoldt. Zur Geschlechtertheorie in abelschen Zahlkörpern. *Math. Nachr.*, **9** (1953), 351–362.

F. Lorenz. Über eine Verallgemeinerung des Hasseschen Normensatzes. *Math. Z.*, **173** (1980), 203–210.

H. Muthsam. Eine Bemerkung über die Wurzelpolynome Galoisscher Gleichungen. *Monatsh. Math.*, **83** (1977), 155–157.

T. Nakayama. Über die Beziehungen zwischen den Faktorensystemen und der Normklassengruppe eines Galoisschen Erweiterungskörpers. *Math. Ann.*, **112** (1936), 85–91.

T. Nakayama. A theorem on the norm group of a finite extension field. *Jap. J. Math.*, **18** (1943), 877–885.

E. Noether. Der Hauptgeschlechtssatz für relative-Galoissche Zahlkörper. *Math. Ann.*, **108** (1933), 411–419.

———. Hyperkomplexe Systeme in ihren Beziehungen zur kommutativen Algebra und Zahlentheorie. *Verhandl. des Internationalen Kongr.*, Zurich 1932, Vol. 1, 189–194.

Y. P. Razmyslov. On a problem of Kaplansky. *Transl. Math. USSR Izv.*, **7** (1973), 479–496.

O. Schilling. Über gewisse Beziehungen zwischen der Arithmetik hyperkomplexer Zahlsysteme und algebraischer Zahlkörper. *Math. Ann.*, **111** (1935), 372–398.

A. Scholz. Totale Normenreste, die keine Normen sind, als Erzeuger nicht abelscher Körpererweiterungen, I. *J. reine angew. Math.*, **172** (1936), 100–107; II, *ibid*, **182** (1940), 217–234.

I. Schur. Über Ringbereiche im Gebiete der ganzzahligen linearen Substitutionen. *Ges. Abh.*, **II**, 359–382.

I. R. Shafarevich. On Galois groups of p-adic fields. *C. R. Dokl. Acad. Sci. URSS*, **53** (1946), 15–16.

H. M. Stark. The genus theory of number fields. *Comm. Pure Appl. Math.*, **29** (1976), 805–811.

H. D. Steckel. Abelsche Erweiterungen mit vorgegebenem Zahlknoten. Preprint.

J. Tate. Global class field theory. In Cassels–Fröhlich (Eds.), *Algebraic Number Theory*, Academic Press: London, 1967.

O. Taussky. On a theorem of Latimer and MacDuffee. *Can. J. Math.*, **1** (1949), 300–302.

———. On matrix classes corresponding to an ideal and its inverse. *Ill. J. Math.*, **1** (1957), 108–113.

———. Ideal matrices, I. *Archiv Math.*, **I** (1962), 275–282.

———. On the similarity transformation between an integral matrix with irreducible characteristic polynomial and its transpose. *Math. Ann.*, **166** (1966), 60–63.

———. Norms from quadratic fields and their relations to non-commuting 2 × 2 matrices. *Monatsh. Math.*, **82** (1972), 253–255.

———. Additive commutators of rational 2 × 2 matrices. *Linear Alg. and Appl.*, **12** (1975), 1–6.

———. From cyclic algebras of quadratic fields to central polynomials. *J. Austral. Math. Soc.*, **25** (1978), 503–506.

———. Composition of binary integral quadratic forms via integral 2 × 2 matrices and composition of matrix classes. *Linear and Multilinear Alg.*, **10** (1981), 309–318.

———. Two facts concerning rational 2 × 2 matrices leading to integral ternary forms representing zero. In *Ternary Quadratic Forms and Norms*. Marcel Dekker: New York, 1982, pp. 39–48.

F. Terada. On the principal genus theorem concerning the abelian extensions. *Tohoku Math. J.*, **4** (1952), 141–152.

A. Weil. Sur la théorie du corps de classes. *J. Math. Soc. Japan*, **3** (1951), 1–35.

A. Weiss. Characteristic polynomials of symmetric matrices. In *Ternary Quadratic Forms and Norms*. O. Taussky (Ed.). Marcel Dekker: New York, 1982, pp. 59–74.

H. Zassenhaus. Neuer Beweis der Endlichkeit der Klassenzahl bei unimodularer Aquivalenz endlicher ganzzahliger Substitutionsgruppen. *Abh. Math. Sem. Univ. Hamburg*, **12** (1938), 276–288.

H. Zassenhaus. A theorem on cyclic algebras. In *Number Theory and Algebra*. Academic Press: New York, 1977, pp. 363–393.

Representations of Lie Groups and the Orbit Method

MICHELE VERGNE*†

It was a great honor and pleasure to be invited by the Association for Women Mathematicians, to give an address for Emmy Noether's 100th birthday.

I have chosen here to talk about my own present mathematical interests, the theory of Lie group representations, and I would like on this occasion to point out that many women are now contributing to the development of this field. This fact would have pleased Emmy Noether. Although she was extraordinary, she would not have thought of herself in these terms, she would have been against holding up her name as a yardstick by which to measure all past, present, and future accomplishments of mathematicians, women and men. We want to celebrate her as an ordinary woman, who could find at this time, only in herself, the necessary courage and inner peace to be what she has been: Emmy Noether.

Let G be a Lie group, i.e. G is an analytic manifold with a group structure such that the group operations are analytic. Let us denote by e the identity element of G.

Consider $\mathfrak{g}$ the tangent space to G at e. For every $X \in \mathfrak{g}$, there exists a homomorphism $\mathbf{R} \to G\colon t \to \exp tX$ such that $(X \cdot \varphi)(e) = (d/dt)\varphi(\exp tX)|_{t=0}$ for every differentiable function φ on G. The map $\exp\colon \mathfrak{g} \to G$ given by $X \to \exp X$ is called the exponential map.

If $g_0 \in G$, the right translation $R(g_0) \cdot g = gg_0$ defines a right action of G on G. If $X \in \mathfrak{g} = T_e(G)$, we denote by X^* the vector field on G such that $(X^*)_{g_0} = R(g_0)_* \cdot X$. The bracket of two elements of $\mathfrak{g}$ is then defined by the relation $[X, Y]^* = [X^*, Y^*]$, where on the right-hand side the bracket is the bracket of vector fields, i.e. $[X^*, Y^*] \cdot \varphi = X^*Y^* \cdot \varphi - Y^*X^* \cdot \varphi$. This defines a Lie algebra structure on $\mathfrak{g}$.

* Department of Mathematics, Massachusetts Institute of Technology, Cambridge, MA 02139, U.S.A.

† I would like to thank the Association for Women Mathematicians and the American Mathematical Society for organizing this event. I also would like to thank Peter Dourmashkin, Devra Garfinkle, Linda Keen, Sophie Koulouras, Lisa Mantini, and Kenneth Manning for comments, help, and positive reinforcements.

Consider the adjoint action of G on itself by inner automorphisms $(\mathrm{Ad}\, g_0) \cdot g = g_0 g g_0^{-1}$. The differential of the map $\mathrm{Ad}\, g_0 \colon G \to G$ at the identity gives rise to a map (still denoted by $\mathrm{Ad}\, g$) on $T_e(G) = \mathfrak{g}$. We have then $g \cdot (\exp tX) g^{-1} = \exp(t \,\mathrm{Ad}\, g \cdot X)$. (We may write $g \cdot X$ for $\mathrm{Ad}\, g \cdot X$.) This action of G on $\mathfrak{g}$ is called the adjoint action. We define also $(\mathrm{ad}\, X) \cdot Y = [X, Y]$, for $X, Y \in \mathfrak{g}$.

I. Examples of Lie Groups

Let us start by giving some examples of Lie groups G. Of course, the most evident example is

1.1. EXAMPLE 1: $G =$ Vector Space. $G = V$ is a finite dimensional real vector space (with the addition law).

Clearly $\mathfrak{g} = V$, and the exponential map is the identity map.

As G is commutative, the adjoint action of G on $\mathfrak{g}$ is trivial, i.e. $g \cdot v = v$ for every $g \in G$, $v \in \mathfrak{g}$.

A closely related example is the example of a torus.

1.2. EXAMPLE 2: $G =$ Torus. Let us consider $G = \{z; z \in \mathbf{C}, |z| = 1\}$ (with the multiplicative law).

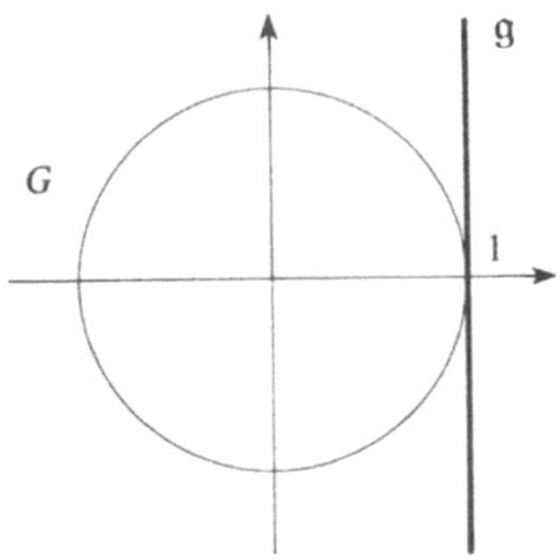

Figure 1

The tangent space $\mathfrak{g}$ at the identity is

$$\mathfrak{g} = i\mathbf{R}.$$

The exponential map is the map $i\theta \to e^{i\theta}$. The adjoint action is trivial.

1.3. More generally, let Γ be a lattice in a vector space V and consider

$$G = T = V/\Gamma.$$

The Lie algebra of G is naturally identified with V.

$$\mathfrak{g} = V.$$

The exponential map $\exp\colon V \to V/\Gamma$ is the natural quotient map.
The adjoint action is trivial.

1.4. Let us now consider the "basic" example of a Lie group:

EXAMPLE 3: The Full Linear Group $GL(n, \mathbf{R})$. Let

$$\begin{aligned} G &= GL(n, \mathbf{R}) \\ &= \{n \times n \text{ invertible real matrices}\}. \end{aligned}$$

As G is an open subset of all $n \times n$ real matrices, we have:

$$\begin{aligned} \mathfrak{g} &= \mathfrak{gl}(n, \mathbf{R}) \\ &= \{n \times n \text{ real matrices}\}. \end{aligned}$$

The exponential map is the usual matrix exponential

$$\exp X = 1 + X + \frac{X^2}{2!} + \cdots + \frac{X^n}{n!} + \cdots.$$

The Lie algebra structure on $\mathfrak{gl}(n, \mathbf{R})$ is $[A, B] = A \circ B - B \circ A$. The adjoint action is given by conjugation:

$$(\mathrm{Ad}\, g) \cdot A = gAg^{-1}.$$

1.5. For the purpose of this discussion, it will be sufficient to consider here linear groups G. A linear group G is a closed subgroup of $GL(n, \mathbf{R})$. Thus its Lie algebra $\mathfrak{g}$ is a subalgebra of $\mathfrak{gl}(n, \mathbf{R})$. The corresponding notions of exponential bracket, and adjoint action are therefore the restrictions to $\mathfrak{g}$ and G of the preceding operations, i.e., for:

$$\mathfrak{g} \subset \mathfrak{gl}(n, \mathbf{R}),$$

$$G \subset GL(n, \mathbf{R}),$$

we still have:

$$\exp X = 1 + X + \frac{X^2}{2!} + \cdots + \frac{X^n}{n!} + \cdots, \qquad X \in \mathfrak{g},$$

$$[A, B] = A \circ B - B \circ A, \qquad A, B \in \mathfrak{g},$$

$$g \cdot X = gXg^{-1}, \qquad g \in G, \quad X \in \mathfrak{g}.$$

Let us consider some examples of such linear groups:

1.6. EXAMPLE 4: The Heisenberg Group. We consider the group

$$G = \left\{ \begin{pmatrix} 1 & x & z \\ 0 & 1 & y \\ 0 & 0 & 1 \end{pmatrix}; x, y, z \in \mathbf{R} \right\}.$$

It is easy to see that:

$$\mathfrak{g} = \left\{ \begin{pmatrix} 0 & p & e \\ 0 & 0 & q \\ 0 & 0 & 0 \end{pmatrix}; p, q, e \in \mathbf{R} \right\}.$$

We write a basis of $\mathfrak{g}$ as follows:

$$P = \begin{pmatrix} 0 & 1 & 0 \\ 0 & 0 & 0 \\ 0 & 0 & 0 \end{pmatrix}, \quad Q = \begin{pmatrix} 0 & 0 & 0 \\ 0 & 0 & 1 \\ 0 & 0 & 0 \end{pmatrix}, \quad E = \begin{pmatrix} 0 & 0 & 1 \\ 0 & 0 & 0 \\ 0 & 0 & 0 \end{pmatrix}.$$

We have the following relations:

$$[P, Q] = P \circ Q - Q \circ P = E,$$
$$[P, E] = 0,$$
$$[Q, E] = 0.$$

1.7. EXAMPLE 5: $G = \mathrm{SU}(2)$. Let us consider the vector space $\mathbf{C}^2$ with its usual Hermitian inner product $|z_1|^2 + |z_2|^2$. We consider the group U(2) of all complex linear transformations g of $\mathbf{C}^2$ leaving this inner product stable. To be able later on to illustrate our notions by pictures in $\mathbf{R}^3$, we will consider the subgroup SU(2) of U(2) of elements g in U(2) such that $\det g = 1$. It is easy to see that:

$$\mathrm{SU}(2) = \left\{ \begin{pmatrix} \alpha & \beta \\ -\bar{\beta} & \bar{\alpha} \end{pmatrix}; \alpha, \beta \in \mathbf{C}, |\alpha|^2 + |\beta|^2 = 1 \right\}.$$

Hence SU(2) is a compact manifold of dimension 3. The Lie algebra $\mathfrak{su}(2)$ of the group SU(2) consists of 2×2 complex matrices X which are anti-hermitian and have trace zero. Thus:

$$\begin{aligned} \mathfrak{g} &= \mathfrak{su}(2) \\ &= \left\{ \begin{pmatrix} ix_3 & -x_1 + ix_2 \\ x_1 + ix_2 & -ix_3 \end{pmatrix}; x_i \in \mathbf{R} \right\}. \end{aligned}$$

We remark here for later use that the function $X \mapsto \det X = x_1^2 + x_2^2 + x_3^2$ is invariant under the adjoint action $g \cdot X = gXg^{-1}$ of G on $\mathfrak{g}$.

1.8. EXAMPLE 6: $\mathrm{SL}(2, \mathbf{R})$. We consider the group of 2×2 real matrices with determinant 1, i.e.

$$G = \left\{ \begin{pmatrix} a & b \\ c & d \end{pmatrix}; a, b, c, d \in \mathbf{R}; ad - bc = 1 \right\}.$$

Thus G is a closed subgroup of $\mathrm{GL}(2, \mathbf{R})$. It is easy to see that:

$$\begin{aligned} \mathfrak{g} &= \mathfrak{sl}(2, \mathbf{R}) = \{\text{matrices of trace zero}\} \\ &= \left\{ X = \begin{pmatrix} x_1 & x_2 + x_3 \\ x_2 - x_3 & -x_1 \end{pmatrix}, x_i \in \mathbf{R} \right\}. \end{aligned}$$

Let us also remark here that the function $X \mapsto \det X = x_3^2 - (x_1^2 + x_2^2)$ is invariant under the adjoint action $g \cdot X = gXg^{-1}$ of G on $\mathfrak{g}$.

II. The Dual of G and the Plancherel Formula

2.1. One of the main objects of representation theory of Lie groups is the study of the dual $\hat{G}$ of the Lie group G or, "equivalently," of the characters of the group G We will now define these notions, allowing some imprecisions from time to time. Let us refer to the useful book [37] for more details.

A representation of G on a vector space V is a homomorphism $g \mapsto T(g)$ of G into the group of linear transformations of V, i.e. we have

$$T(g_1 g_2) = T(g_1) \circ T(g_2),$$
$$T(e) = \mathrm{id}_V.$$

The representation T of G in V is called a unitary representation of G, if V is a Hilbert space, the operators $T(g)$ unitary operators and the map $(g, v) \mapsto T(g) \cdot v$ continuous.

We have an obvious notion of equivalence of unitary representations. Two unitary representations T_1, T_2 of G in Hilbert spaces H_1, H_2 are equivalent if there exists a unitary isomorphism I: $H_1 \to H_2$ such that the following diagram:

$$\begin{array}{ccc} H_1 & \xrightarrow{I} & H_2 \\ \uparrow{\scriptstyle T_1(g)} & & \uparrow{\scriptstyle T_2(g)} \\ H_1 & \xrightarrow{I} & H_2 \end{array}$$

is commutative, for every $g \in G$.

If (T_1, H_1) and (T_2, H_2) are two unitary representations of G, we can form the representation $T = T_1 \oplus T_2$ acting on the direct sum $H = H_1 \oplus H_2$ by $T(g) = T_1(g) \oplus T_2(g)$. A representation T is irreducible if T is not obtained as a direct sum of two representations. Equivalently, (T, H) is irreducible if there exists no proper Hilbert subspace of H invariant under T. As every unitary representation T of G in a Hilbert space H is a "sum" (eventually a "continuous sum") of unitary irreducible representations of G, the essential objects of unitary representation theory are the irreducible representations of G.

By definition, the dual $\hat{G}$ of G is the set of equivalence classes of unitary irreducible representations of G.

2.2. Let us give now an important example of a unitary representation: the regular representation. Consider on G the left invariant Haar measure dg (unique up to a positive scalar multiple). Consider the Hilbert space $L^2(G)$ of dg-square integrable functions on G, i.e.

$$L^2(G) = \left\{ f;\ \int |f(g)|^2\, dg < \infty \right\}.$$

The left action $(L(g_0)f)(g) = f(g_0^{-1}g)$ defines a unitary representation of G on $L^2(G)$. This representation is highly reducible, and an important problem is to describe explicitly the decomposition of this representation into irreducible representations.

2.3. Let us now define the notion of the character of a representation. If T is a representation of G in a finite dimensional vector space V, the character of T is the function $\chi_T(g) = \operatorname{tr} T(g)$. It is clear that this function is invariant under conjugation, i.e. $\chi_T(g_0 g g_0^{-1}) = \chi_T(g)$. It is well known that if G is a compact group, the description of $\hat{G}$ is equivalent to the description of all the functions χ_T.

Let us now consider the case of an infinite dimensional representation. Define, for a function φ in $L^1(G)$, the operator $T(\varphi) = \int_G \varphi(g)T(g)\,dg$, i.e.

$$\langle T(\varphi)x, y\rangle = \int_G \varphi(g)\langle T(g)x, y\rangle\,dg.$$

If (T, H) is irreducible and if φ is a C^∞ function with compact support on G, it is often the case (in particular for all our examples) that the operator $T(\varphi)$ has a trace. If e_i is any orthonormal basis of H, we have $\operatorname{tr} T(\varphi) = \sum_i \langle T(\varphi)e_i, e_i\rangle$. Furthermore, the map $\varphi \to \operatorname{tr} T(\varphi)$ happens to be a distribution on G. This distribution (if defined) is called the character of the representation T. The distribution character $\operatorname{tr} T$ depends of the choice of dg. It is clear that, if T is a representation of G in a finite dimensional vector space V, we have

$$\operatorname{tr} T(\varphi) = \int_G (\operatorname{tr} T(g))\varphi(g)\,dg.$$

2.4. A group G is called unimodular if the left Haar measure is also right invariant. Equivalently, G is unimodular, if $|\det \operatorname{Ad} g| = 1$ for every $g \in G$. If G is unimodular, the distribution $\operatorname{tr} T$ (if defined) is an invariant (by the adjoint action) distribution on G, as:

$$\begin{aligned}(\operatorname{tr} T, (\operatorname{Ad} g_0)^{-1} \cdot \varphi) &= \operatorname{tr} \int_G T(g)\varphi(g_0 g g_0^{-1})\,dg \\ &= \operatorname{tr} \int_G T(g_0^{-1} g g_0)\varphi(g)\,dg \\ &\qquad \text{as } dg \text{ is left and right invariant,} \\ &= \operatorname{tr} \int_G T(g_0)^{-1}T(g)T(g_0)\varphi(g)\,dg \\ &= \operatorname{tr} T(g_0)^{-1}T(\varphi)T(g_0) \\ &= \operatorname{tr} T(\varphi).\end{aligned}$$

Consider the δ-distribution $\varphi \mapsto \varphi(e)$. It is an invariant distribution on G. The purpose of the Plancherel formula is to express the value of a function φ at e in function of the values of the characters $\operatorname{tr} T(\varphi)$. It is indeed possible to do so, for any "tame" unimodular Lie group (we will consider here only real algebraic groups, which are "tame" Lie groups), as we have the "abstract Plancherel theorem":

2.5. Theorem (see [14]). *Let G be a tame unimodular group. There exists a measure $d\mu$ (called the Plancherel measure) on $\hat{G}$ such that:*

$$\varphi(e) = \int_{\hat{G}} \operatorname{tr} T(\varphi)\, d\mu(T)$$

for every C^∞ function φ on G with compact support.

Let us first remark here that the support of the measure $d\mu$ may be smaller than $\hat{G}$. This already happens for the group SL(2, **R**). We will denote by $\hat{G}_r$ the support of $d\mu$ and call $\hat{G}_r$ the reduced dual of G.

2.6. Let us come back to our examples and describe accordingly the dual $\hat{G}$ of these groups and the Plancherel formula.

2.7. Example 1: $G = V$.

(a) *The set* $\hat{G}$

Let V be a real finite dimensional vector space and V^* be the dual vector space. If $f \in V^*$, we consider $\chi_f(x) = e^{i(f,x)}$. The map $x \to \chi_f(x)$ defines a character of the additive group V. (We also use the word character for a 1-dimensional representation of a group G.) Thus we have:

$$\hat{V} \simeq V^*.$$

(b) *Characters*

Let dx be a Lebesgue measure on V. If φ is a C^∞ function with compact support, we have

$$\operatorname{tr} \chi_f(\varphi) = \int_V \varphi(x)\chi_f(x)\, dx = \int_V \varphi(x)e^{i(f,x)}\, dx = \hat{\varphi}(f).$$

Thus $f \mapsto \operatorname{tr} \chi_f(\varphi)$ is the function on V^* given by the Fourier transform of φ.

(c) *The Plancherel formula*

The Plancherel formula is the usual Plancherel inversion formula:

$$\varphi(0) = \int_{V^*} \hat{\varphi}(f)\, df,$$

where df is the dual Lebesgue measure on $V^* = \hat{V}$.

2.8. Example 2: $G = T$.

(a) *The set* $\hat{G}$

Let $G = T = \{z; z \in \mathbf{C}, |z| = 1\}$.

If $n \in \mathbf{Z}$, define $\chi_n(z) = z^n$. We have

$$\hat{G} \simeq \mathbf{Z} \qquad (n \leftrightarrow \chi_n).$$

(b) *Characters*

Let us choose on T the Haar measure giving total mass 1 to T.

If φ is a C^∞ function on T, then

$$\operatorname{tr} \chi_n(\varphi) = \int \varphi(e^{i\theta}) e^{in\theta} \frac{d\theta}{2\pi}$$

is the $(-n)$th Fourier coefficient of the periodic function $\theta \to \varphi(e^{i\theta}) = \sum a_n e^{in\theta}$.

(c) *The Plancherel formula*

The Plancherel formula is deduced immediately from the expansion of a function in its Fourier coefficients:

$$\varphi(1) = \sum \operatorname{tr} \chi_n(\varphi).$$

2.9. More generally, let $T = V/\Gamma$ be a n-dimensional torus. It is clear that every representation of T gives, by composition with the natural projection $V \xrightarrow{\exp} V/\Gamma = T$, a representation of V. Thus $\hat{T}$ is included in V^*.

Figure 2

This map is represented for $V = \mathbf{R}$, $\Gamma = 2\pi\mathbf{Z}$, by:

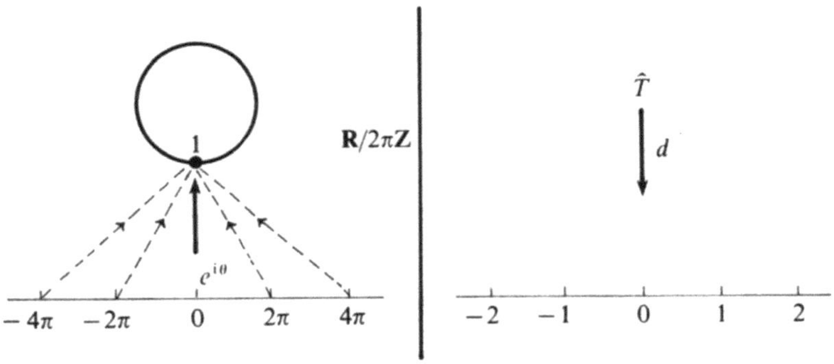

Figure 3

It is obvious that a character χ_f of V gives rise to a character of T, if and only if $\chi_f(x) = e^{i(f,x)} = 1$, for every x in V such that $\exp x = 1$, i.e. for every x in Γ. Thus we have identified $d(\hat{T})$ with the dual lattice Γ^* of Γ in V^*. Therefore,

(a) $$\hat{T} = \Gamma^* = \{f \in V^*; (f, \gamma) \in 2\pi\mathbf{Z}, \text{ for every } \gamma \in \Gamma\}.$$

Let us normalize the Lebesgue measure on V such that the volume of the fundamental parallelepiped spanned by Γ is 1. Thus we have, for φ a C^∞ function on V/Γ:

(b) $$\operatorname{tr} \chi_f(\varphi) = \int_{V/\Gamma} \varphi(x) e^{i(f,x)}\, dx, \text{ for } f \in \Gamma^*.$$

The Plancherel formula is:

(c) $$\varphi(0) = \sum_{f \in \Gamma^*} \operatorname{tr} \chi_f(\varphi)$$

for every C^∞ function φ on V/Γ.

Let us point out at this occasion that, by an immediate averaging process, this formula is equivalent to:

(d) The Poisson formula:

$$\sum_{\gamma \in \Gamma} \varphi(\gamma) = \sum_{f \in \Gamma^*} \hat{\varphi}(f)$$

for every C^∞ function φ on V with compact support (or in the Schwartz space of V).

2.10. Example 3: $G = \mathrm{GL}(n, \mathbf{R})$. Let us now consider our basic example of a Lie Group $G = \mathrm{GL}(n, \mathbf{R})$. Now is the time to reveal the truth. The dual $\hat{G}$ of $\mathrm{GL}(n, \mathbf{R})$ is not known. The case $n = 2$, the first work in the representation theory of semi-simple groups, was completed by Valentine Bargmann in 1946 [6]. But since this time the general case seems still out of reach.

2.11. Let us mention some of the recent results on this question of the determination of the unitary dual of real semi-simple Lie groups:

If $n \leq 4$, the dual $\hat{G}$ of $G = \mathrm{GL}(n, \mathbf{R})$ has been determined by Birgit Speh [60].

The dual $\hat{G}$ of any complex semi-simple Lie group of rank two (i.e. $\mathrm{SL}(3, \mathbf{C})$, $\mathrm{Sp}(2, \mathbf{C})$, G_2) has been determined by Michel Duflo [16].

The dual $\hat{G}$ of any semi-simple Lie group of real rank one (i.e. $G = \mathrm{SO}(n, 1)$, $\mathrm{SU}(n, 1)$, $\mathrm{Sp}(n, 1)$ or a real form of F_4) has been determined by Welleda Baldoni-Silva and Dan Barbasch [4], [5].

The dual $\hat{G}$ of $G = \mathrm{SU}(2, 2)$ has been determined by Anthony Knapp and Birgit Speh [38].

Thus we still know very little about unitary representation theory of real semi-simple Lie groups. However, as remarked before, the Plancherel inversion formula for a Lie group does not involve the complete description of $\hat{G}$, but only of the reduced dual $\hat{G}_r$. If G is a real semi-simple Lie group (with finite center), the reduced dual $\hat{G}_r$ of $\hat{G}$ as well as the "concrete" Plancherel measure $d\mu(T)$ on $\hat{G}_r$ has been determined by Harish-Chandra [26]. We will discuss further some of the corresponding results.

2.12. Example 4: G = The Heisenberg Group.

(a) *The set* $\hat{G}$

The description of the dual $\hat{G}$ of the Heisenberg group was the object of the famous theorem of Stone–Von Neumann [1931] on the "Uniqueness for the Schrödinger operators" [47].

Here is the complete list of all equivalence classes of unitary irreducible representations of G.

(1) Let $H = L^2(\mathbf{R})$.

For each λ a non-zero real number, consider the representation T_λ of G in $L^2(\mathbf{R})$ given by:

$$(T_\lambda \exp(tP) \cdot \varphi)(y) = e^{i\lambda t y}\varphi(y),$$
$$(T_\lambda \exp(tQ) \cdot \varphi)(y) = \varphi(y - t),$$
$$(T_\lambda \exp(tE) \cdot \varphi)(y) = e^{i\lambda t}\varphi(y) \quad \text{for } \varphi \in L^2(\mathbf{R}).$$

(2) Let $H = \mathbf{C}$

For each (α, β) a pair of real numbers, consider the character

$$T_{\alpha,\beta}(\exp xP \exp yQ \exp zE) = e^{i\alpha x}e^{i\beta y}.$$

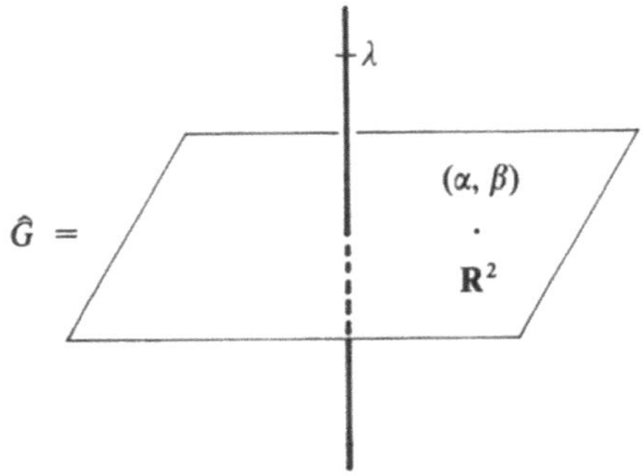

Figure 4

Remark. Differentiating formally the representation T_λ, we obtain

$$dT_\lambda(P) = \frac{d}{dt} T_\lambda \exp(tP)|_{t=0} = i\lambda y,$$

$$dT_\lambda(Q) = \frac{d}{dt} T_\lambda \exp(tQ)|_{t=0} = -\frac{\partial}{\partial y},$$

$$dT_\lambda(E) = \frac{d}{dt} T_\lambda \exp(tE)|_{t=0} = i\lambda.$$

The operators $i\lambda y$, $-\partial/\partial y$ determine a representation of the "Schrodinger operators" by skew-adjoint operators, in particular these operators satisfy the canonical commutation relation: $P \circ Q - Q \circ P = i\lambda \text{ Id}$.

(b) *Characters*
Let us determine the characters of the corresponding representations. It is not difficult to compute that, for φ a C^∞ function on G with compact support:

$$\operatorname{tr} T_\lambda(\varphi) = \frac{2\pi}{\lambda} \int_{\mathbf{R}} \varphi(\exp zE) e^{i\lambda z}\, dz,$$

with the Haar measure on G given by $dx\, dy\, dz$.

(c) *The Plancherel formula*
We have the Plancherel formula

$$\varphi(e) = \int_{\mathbf{R}} \operatorname{tr} T_\lambda(\varphi) \frac{\lambda\, d\lambda}{(2\pi)^2}.$$

Proof. From the formula given in (b) for the characters, the Plancherel formula is immediately deduced from the usual Fourier inversion formula.

2.13. Example 5: $G = \mathrm{SU}(2)$.

(a) *The set* $\hat{G}$
Let n be a positive integer and consider:

$$V_n = \{\text{Polynomials on } \mathbf{C}^2 \text{ of homogeneous degree } n - 1\}$$
(the dimension of the vector space V_n is n).

The group SU(2) acts on $\mathbf{C}^2$, and thus acts on polynomials on $\mathbf{C}^2$ via $(g \cdot P)(z) = P(g^{-1} \cdot z)$. It is clear that this action preserves the space V_n. We have

$$\hat{G} \simeq N$$

by $[n \leftrightarrow (T_n(g)P)(z) = P(g^{-1}z),\, P \in V_n]$.

Figure 5

(b) *Characters*
Let $\begin{pmatrix} e^{i\theta} & 0 \\ 0 & e^{-i\theta} \end{pmatrix}$ be an element of SU(2). The polynomials $z_1^i z_2^j$ $(i + j = n - 1)$ are eigenvectors for the transformation $T_n\begin{pmatrix} e^{i\theta} & 0 \\ 0 & e^{-i\theta} \end{pmatrix}$. Thus we can easily compute:

$$\operatorname{tr} T_n\begin{pmatrix} e^{i\theta} & 0 \\ 0 & e^{-i\theta} \end{pmatrix} = \sum_{k=0}^{n-1} e^{i(n-1-2k)\theta} = \frac{e^{in\theta} - e^{-in\theta}}{e^{i\theta} - e^{-i\theta}}.$$

(This formula is a special case of the Weyl formula for characters [64].)

(c) *The Plancherel formula*
We have the Plancherel formula

$$\varphi(e) = \sum_{n>0} n \operatorname{tr} T_n(\varphi).$$

A proof of this will be given in Appendix 2.

2.14. EXAMPLE 6: $G = \mathrm{SL}(2, \mathbf{R})$.

(a) *The reduced dual* $\hat{G}_r$
The set $\hat{G}$ has been determined in the classical article of V. Bargmann [6], still the best reference on this subject. The complete list of the unitary irreducible representations of $\mathrm{SL}(2, \mathbf{R})$ consists of "discrete series," "principal series," and "complementary series." Only the first two series contribute to the Plancherel measure. Furthermore, the description of the complementary series is more subtle, so we will list here only the set $\hat{G}_r$. It consists of:

(1) The discrete series
(1.1) Let n be a positive integer. Let us consider the upper half-plane $\mathbf{P}^+ = \{z = x + iy; x, y \in \mathbf{R}, y > 0\}$. The group $\mathrm{SL}(2, \mathbf{R})$ acts as a group of holomorphic transformations on $\mathbf{P}^+$ by $z \mapsto g \cdot z = (az + b)/(cz + d)$.
Let n be a positive integer. Consider

$$H_n = \left\{\varphi, \text{ holomorphic functions on } \mathbf{P}^+ \text{ such that } \int_{\mathbf{P}^+} |\varphi|^2 y^{n-1}\, dx\, dy < \infty\right\}.$$

(This Hilbert space would be $\{0\}$, if n were negative.)
For

$$g = \begin{pmatrix} a & b \\ c & d \end{pmatrix} \in \mathrm{SL}(2, \mathbf{R}),$$

define

$$(T_n(g^{-1})\varphi)(z) = (cz + d)^{-(n+1)}\varphi\left(\frac{az + b}{cz + d}\right).$$

The map $g \mapsto T_n(g)$ defines a unitary irreducible representation of G in H_n.
The series of representations (T_n, H_n) $(n \geq 1)$ is called the holomorphic discrete series.

(1.2) Let n be a negative integer. Consider

$$H_n = \left\{\varphi, \text{ antiholomorphic functions on } \mathbf{P}^+ \text{ such that } \int_{\mathbf{P}^+} |\varphi|^2 y^{|n|-1}\, dx\, dy < \infty\right\}.$$

Then, for

$$g = \begin{pmatrix} a & b \\ c & d \end{pmatrix} \in \mathrm{SL}(2, \mathbf{R}),$$

consider

$$(T_n(g^{-1})\varphi)(z) = \overline{(cz + d)}^{-(1-n)}\varphi\left(\frac{az + b}{cz + d}\right).$$

The map $g \mapsto T_n(g)$ defines a unitary irreducible representation of G in H_n.

The series of representations (T_n, H_n) $(n \leq -1)$ is called the antiholomorphic discrete series.

(2) The principal series

Let $H = L^2(\mathbf{R})$, let s be a non-negative real number. Define the representations $T_s^{\pm}$ of G in $L^2(\mathbf{R})$ by:

$$(T_s^+(g^{-1})f)(x) = |cx + d|^{-1+is} f\left(\frac{ax + b}{cx + d}\right),$$

$$(T_s^-(g^{-1})f)(x) = \operatorname{sign}(cx + d)|cx + d|^{-1+is} f\left(\frac{ax + b}{cx + d}\right), \quad \text{if } g = \begin{pmatrix} a & b \\ c & d \end{pmatrix}.$$

It is easy to verify that T_s^- and T_s^- are unitary representations of G in $L^2(\mathbf{R})$. Furthermore, T_s^+ and T_s^- are irreducible, except for the representation T_0^- which breaks up into two irreducible pieces. The series of representations (T_s^+, T_s^-) for $s \geq 0$ forms the two principal series of SL(2, **R**).

A schematic diagram of $\hat{G}_r$ for $G = \mathrm{SL}(2, \mathbf{R})$ is thus:

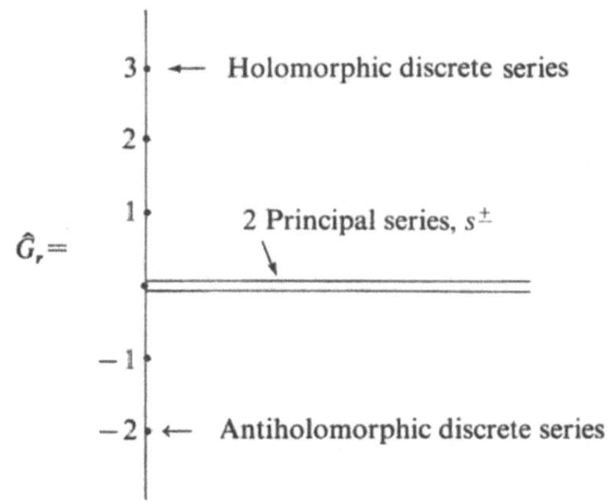

Figure 6

(b) *Characters*

We give here the results on the characters of the representations of SL(2, **R**) as proven by Harish-Chandra [24]. First of all, as the case for any real semi-simple

Lie group G, the character distribution $\varphi \to \operatorname{tr} T(\varphi)$ is well defined for every T in $\hat{G}$ and is given by integration against a locally L^1-function Θ_T, i.e.

$$\operatorname{tr}\left(\int_G T(g)\varphi(g)\,dg\right) = \int_G \Theta_T(g)\varphi(g)\,dg$$

for φ a C^∞ function on G with compact support.

Clearly $\Theta_T(g_0 g g_0^{-1}) = \Theta_T(g)$. Thus Θ_T is determined by its restriction to the subsets

$$B = \left\{\begin{pmatrix} \cos\theta & \sin\theta \\ -\sin\theta & \cos\theta \end{pmatrix}; \theta \in \mathbf{R}\right\},$$

and

$$A = \left\{\begin{pmatrix} a & 0 \\ 0 & a^{-1} \end{pmatrix}; a \in \mathbf{R}^*\right\},$$

as almost all (for dg) elements of SL(2, $\mathbf{R}$) are conjugated to an element of one of these two sets [SL(2, $\mathbf{R}$) has two conjugacy classes of Cartan subgroups].

We have the following formulae:

(1) Discrete series

(1.1) Holomorphic discrete series $n \geq 1$

$$\Theta_{T_n}\begin{pmatrix} \cos\theta & -\sin\theta \\ \sin\theta & \cos\theta \end{pmatrix} = \frac{e^{in\theta}}{e^{-i\theta} - e^{i\theta}},$$

$$\Theta_{T_n}\begin{pmatrix} \varepsilon e^t & 0 \\ 0 & \varepsilon e^{-t} \end{pmatrix} = \varepsilon^{n+1}\frac{e^{-|nt|}}{|e^t - e^{-t}|} \quad \text{if } \varepsilon = \pm 1.$$

(1.2) Antiholomorphic discrete series $n \leq -1$

$$\Theta_{T_n}\begin{pmatrix} \cos\theta & -\sin\theta \\ \sin\theta & \cos\theta \end{pmatrix} = \frac{e^{in\theta}}{e^{i\theta} - e^{-i\theta}},$$

$$\Theta_{T_n}\begin{pmatrix} \varepsilon e^t & 0 \\ 0 & \varepsilon e^{-t} \end{pmatrix} = \varepsilon^{1-n}\frac{e^{-|nt|}}{|e^t - e^{-t}|}.$$

(2) Principal series

$$\Theta_{T_s^\pm}\begin{pmatrix} \cos\theta & -\sin\theta \\ \sin\theta & \cos\theta \end{pmatrix} = 0,$$

$$\Theta_{T_{s^+}}\begin{pmatrix} \varepsilon e^t & 0 \\ 0 & \varepsilon e^{-t} \end{pmatrix} = \frac{e^{ist} + e^{-ist}}{|e^t - e^{-t}|},$$

$$\Theta_{T_{s^-}}\begin{pmatrix} \varepsilon e^t & 0 \\ 0 & \varepsilon e^{-t} \end{pmatrix} = \varepsilon\frac{e^{ist} + e^{-ist}}{|e^t - e^{-t}|}.$$

(c) *The Plancherel formula for G*
We have the formula;

$$2\pi\varphi(e) = \sum_{\substack{n\in\mathbf{Z}\\ n\neq 0}} |n| \operatorname{tr} T_n(\varphi)$$

$$+\tfrac{1}{2}\int_{\mathbf{R}^+} s \tanh\frac{\pi s}{2} \operatorname{tr} T_s^+(\varphi)\, ds$$

$$+\tfrac{1}{2}\int_{\mathbf{R}^+} s \coth\frac{\pi s}{2} \operatorname{tr} T_s^-(\varphi)\, ds,$$

which can be deduced from the preceding formula for the characters (see [41]).

2.15. Let G be a general real semi-simple Lie group. We give now a cursory summary of results on unitary representation theory of G. As we pointed out before, only the part $\hat{G}_r$ of $\hat{G}$ is known, while a complete description of $\hat{G}$ is still an unsolved problem. Let us, for example, mention that a most remarkable representation of the symplectic group, the Shale–Weil representation, is a singular unitary representation (singular in the sense that its two components are not in $\hat{G}_r$) and that its existence has not been recaptured by any systematic procedure.

Let us center our attention on $\hat{G}_r$ and the Plancherel formula. The central reference for this topic is the work of Harish-Chandra. As in the case of SL(2, $\mathbf{R}$), we may list representations in $\hat{G}_r$ by series. Let Car G be the set of conjugacy classes of Cartan subgroups of G. There is as many series as elements in Car G: to a conjugacy class of a Cartan subgroup H of G corresponds a series of representations $\{T_i, i \in I_H\}$. The elements of this series may be indexed by a subset I_H of $\hat{H}$. [I_H parametrizes the set of regular characters of H modulo the action of a finite group. For example, in the case of SL(2, $\mathbf{R}$), there are two conjugacy classes of Cartan subgroups, namely the conjugacy class of B and the one of A. The discrete series is indexed by the set of characters of B, except the trivial character, i.e. by $\mathbf{Z} - \{0\}$. The principal series is indexed by $\hat{A} = \mathbf{Z}/2\mathbf{Z} \times \mathbf{R}$ modulo the action $(\varepsilon, s) \to (\varepsilon, -s)$.]

In general, if G has a compact Cartan subgroup B, the corresponding series $\{T_\lambda, \lambda \in \hat{B}, \lambda \text{ regular}\}$, is indexed by a discrete set. The corresponding representation T_λ whose existence was proven "abstractly" by Harish-Chandra occurs as a discrete summand in the regular representation of G in $L^2(G)$. Thus this series is called the discrete series of G. Let T_λ be a representation of the discrete series. Harish-Chandra gave an explicit formula [25] for the character $\Theta_\lambda(g)\, dg$ of T_λ, in the case g is a regular elliptic element of G. This formula is a finite sum over the fixed points for the action of g on G/B and is formally similar to the Atiyah–Bott fixed point formula [1] for the twisted Dirac operator. It was a remarkable result of W. Schmid [52] that, indeed the representation T_λ can be realized in the space of L^2-solutions of the twisted Dirac operator D_λ on G/B.

The other series $\{T_i; i \in I_H\}$ of representations of G are constructed in a simple way from representations of discrete series of reductive subgroups of G.

The Plancherel measure is thus the measure $d\mu(T)$ on the set $\hat{G}_r = \bigcup_{H \in \mathrm{Car}\,G} I_H$, such that

2.16 $$\varphi(e) = \int_{\hat{G}_r} \mathrm{tr}\; T(\varphi)\, d\mu(T).$$

The explicit formula for $d\mu(T)$ has been determined by Harish-Chandra, and the Plancherel formula proven [26]. May I confess that I never understood the original proof of Harish-Chandra and that I am very grateful to Rebecca Herb to have given recently a more accessible proof?

2.17. For this proof and on its own right, the integrand tr $T(\varphi)$ is worth detailing. It is a difficult question and there have been several attempts to find explicit formulae for it. In an article on this subject [53], W. Schmid went so far as to declare: "For a general group G, it will be very difficult to express the discrete series characters by a completely explicit global formula in closed form—if it can be done at all." But, one should never give up hope and recently Rebecca Herb [27], [31] gave formulae for the locally L^1-function Θ_T defining the distribution character $\Theta_T(g)\, dg$ of a representation T of $\hat{G}_r$. Some of the ingredients for her formulae are related to the work of Diana Shelstad on "Orbital integrals and base change" [54], [57].

When having explicit formulae for the integrand tr $T(\varphi)$ and the Harish-Chandra formula for $d\mu(T)$, it was then (theoretically) simple to reprove 2.17. This was also accomplished by Rebecca Herb for linear semi-simple Lie groups [31], [32].

2.18. We would like now to discuss representation theory of general Lie groups. We will try to give a glimpse on some beautiful and deep results on general Lie groups and show how a large part of the specific results we have described here fit in the general theory of the "orbit method." However, as shown exemplarily by the case of $G = \mathrm{GL}(n, \mathbf{R})$, it would be too much to hope that a single mode of explanation will lead to a total understanding of $\hat{G}$.

III. The Orbit Method

Let us consider a general Lie group G. What kind of parameters should we look for to describe $\hat{G}$? It was A. A. Kirillov [35], who discovered universal parameters for $\hat{G}$, whatever the Lie group G is. This idea is very simple and is referred to as the "orbit method."

3.1. Let G be a general Lie group, $\mathfrak{g}$ the Lie algebra of G and $\mathfrak{g}^*$ the dual vector space of $\mathfrak{g}$. As G operates on $\mathfrak{g}$ by the adjoint action, G operates on $\mathfrak{g}^*$ in such a way that

$$\langle g \cdot f, g \cdot X \rangle = \langle f, X \rangle, \qquad f \in \mathfrak{g}^*, \quad X \in \mathfrak{g}.$$

This action of G in $\mathfrak{g}^*$ is called the coadjoint action. Let us consider the orbits of G in $\mathfrak{g}^*$ under the coadjoint action of G. Kirillov's idea is that the dual $\hat{G}$ of G should be related to the dual vector space $\mathfrak{g}^*$ of $\mathfrak{g}$, or more exactly, related to the set $\mathfrak{g}^*/G$ of orbits of G in $\mathfrak{g}^*$. Let us quote here the first striking result of Kirillov [35].

3.2. Theorem. *Let G be a simply connected nilpotent Lie group, then $\hat{G}$ is isomorphic to $\mathfrak{g}^*/G$.*

I would like to comment here on this theorem. Once the principle of the correspondence is stated (the irreducible representations associated to orbits are constructed by induction) the proof follows in a straightforward manner from G. W. Mackey theory [43]. However, it is the statement itself which is remarkable. This idea generated many new insights on many aspects of representation theory of Lie groups and Lie algebras.

Let us first take a look at the space of orbits for our examples of nilpotent groups, i.e. Examples 1, 2, and 4. (If G is a connected nilpotent Lie group, with universal covering $\tilde{G}$, the set $\hat{G}$ is a subset of $\hat{\tilde{G}}$. Thus $\hat{G}$ is identified with a subset of G-orbits in $\mathfrak{g}^*$, depending on G.)

3.3. Example 1: $G =$ Vector Space. Let $G = V$.
Then

$$\mathfrak{g} = V,$$
$$\mathfrak{g}^* = V^* \simeq \mathfrak{g}^*/G \simeq \hat{G}.$$

Thus the theorem of Kirillov is true in this case. (Fortunately!)

3.4. Example 2: $G = T$. Let $G = T = V/\Gamma$.
Then

$$\mathfrak{g} = V,$$
$$\mathfrak{g}^* = V^* = \mathfrak{g}^*/G,$$
$$\hat{G} \simeq \Gamma^* \subset \mathfrak{g}^*.$$

3.5. Example 3: The Heisenberg Group. Recall that

$$G = \left\{\begin{pmatrix} 1 & x & z \\ 0 & 1 & y \\ 0 & 0 & 1 \end{pmatrix}; x, y, z \in \mathbf{R}\right\},$$

$$\mathfrak{g} = \left\{\begin{pmatrix} 0 & p & e \\ 0 & 0 & q \\ 0 & 0 & 0 \end{pmatrix}; p, q, e \in \mathbf{R}\right\},$$

with basis P, Q, E.

It is easy to compute the adjoint action of G on $\mathfrak{g}$:

$$\begin{pmatrix} 1 & x & z \\ 0 & 1 & y \\ 0 & 0 & 1 \end{pmatrix}\begin{pmatrix} 0 & p & e \\ 0 & 0 & q \\ 0 & 0 & 0 \end{pmatrix}\begin{pmatrix} 1 & x & z \\ 0 & 1 & y \\ 0 & 0 & 1 \end{pmatrix}^{-1} = \begin{pmatrix} 0 & p & e + qx - py \\ 0 & 0 & q \\ 0 & 0 & 0 \end{pmatrix}.$$

The coadjoint action of G in $\mathfrak{g}^*$ is:

$$\begin{pmatrix} 1 & x & z \\ 0 & 1 & y \\ 0 & 0 & 1 \end{pmatrix}^{-1} \cdot (\alpha P^* + \beta Q^* + \lambda E^*) = (\alpha - \lambda y)P^* + (\beta + \lambda x)Q^* + \lambda E^*.$$

Thus we see that the orbit of the point λE^* for $\lambda \neq 0$ is the 2-dimensional plane defined by $(f, E) = \lambda$, while the points $\alpha P^* + \beta Q^*$ are 0-dimensional orbits. We obtain the following pictures of $\mathfrak{g}^*$ and $\mathfrak{g}^*/G$:

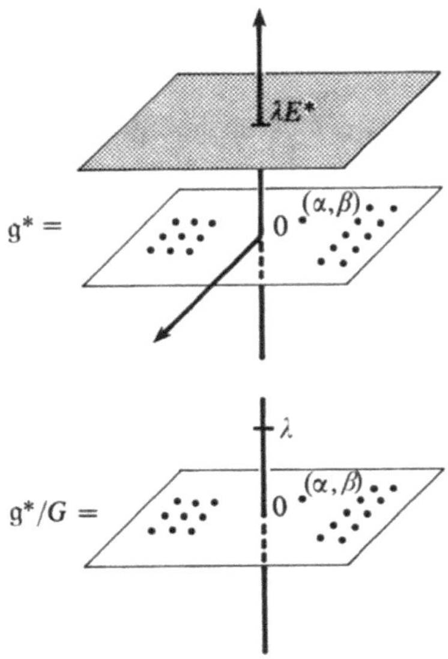

Figure 7

Note that the element $f = \alpha P^* + \beta Q^*$ of $\mathfrak{g}^*$ is a 1-dimensional character of the Lie algebra $\mathfrak{g}$ (i.e. $f[X, Y] = 0$, for every $X, Y \in \mathfrak{g}$). As expected, the Kirillov correspondence assigns to this point f of $\mathfrak{g}^*$ the character of G given by $T_f(\exp X) = e^{i(f, X)}$ which is the representation $T_{\alpha, \beta}$ of Section II, Example 4. The Kirillov correspondence assigns to the orbit of the point λE^* the representation T_λ. Thus we see clearly by comparison of 2.12, Figure 3 and 3.5, Figure 6, how $\mathfrak{g}^*/G$ becomes a natural set of parameters for $\hat{G}$.

3.6. This striking result for nilpotent groups leads to the following question. Is there also for any general Lie group G a relation between $\hat{G}$ and $\mathfrak{g}^*/G$? The answer is "yes." An explanation of the relation calls to my mind many diverse thoughts including analogies, conjectures, and expectations, and it is not possible to give such a simple answer in the general case as in the nilpotent case. I will select here the analytic aspect of these relations given by the Kirillov universal character formula, leaving out many other equally rich aspects, such as the algebraic or geometric ones. For example, the power of Kirillov's orbit method for the study of enveloping algebras was foreshadowed by the work of J. Dixmier [13]. See the book [15] by Dixmier and, among other recent articles, those of Colette Moeglin [44], [45] and Moeglin and Rentschler [46].

3.7. Let $\mathfrak{g}$ be a Lie algebra. Consider the analytic function

$$j(X) = \det\left(\frac{e^{\operatorname{ad} X/2} - e^{-\operatorname{ad} X/2}}{\operatorname{ad} X}\right)$$

on the Lie algebra $\mathfrak{g}$, and define $j^{1/2}$ the analytic square root of j (defined at least in a neighborhood of 0).

Kirillov [36] conjectured the following universal formula for characters: Let G be a tame unimodular Lie group. For almost every representation T in $\hat{G}_r$, there exists an orbit $\mathcal{O}_T$ (of maximal dimension) of G in $\mathfrak{g}^*$ such that we have the equality of generalized functions (at least in a neighborhood of 0),

$$\operatorname{tr} T(\exp X) j^{1/2}(X) = \int_{\mathcal{O}_T} e^{i\langle \xi, X\rangle}\, d\mu(\xi),$$

where $d\mu(\xi)$ is a G-invariant measure on $\mathcal{O}$. (The normalization of $d\mu(\xi)$ will be made precise later on.)

For this equality to hold, we should have:

$$\operatorname{tr} \int_{\mathfrak{g}} T(\exp X) j^{1/2}(X)\varphi(X)\, dX = \int_{\mathcal{O}_T} \left(\int_{\mathfrak{g}} e^{i\langle \xi, X\rangle}\varphi(X)\, dX\right) d\mu(\xi)$$

at least, for every C^∞ function φ on $\mathfrak{g}$ supported in a small neighborhood of 0.

3.8. The Kirillov character formula defines conjecturally a map $T \to \mathcal{O}_T$ from $\hat{G}_r$ to $\mathfrak{g}^*/G$. Of course, in Example 1: $G = V$, the map $f \in V^* \to \chi_f(x) = e^{i(f,x)}$ is also the one compatible with this universal character formula.

What is the image of $\hat{G}_r$ under this map? As it is obvious from the case of compact groups ($\hat{G}$ being a discrete set), not every orbit corresponds to a representation of G. The corresponding orbit should satisfy some integrality conditions, which appear naturally when considering the inverse problem: How to construct from an orbit $\mathcal{O}$ of the coadjoint representation a "natural" representation $T_{\mathcal{O}}$ associated to $\mathcal{O}$ having the prescribed character formula. This inverse problem is referred to as the "quantization" of an orbit and, from my point of view, has no completely satisfactory answer. The reader should consult the book [23] by Guillemin and Sternberg for insights on possible "quantization methods."

3.9. Let us now discuss these integrability conditions on $\mathcal{O}$, which arise already naturally in a preliminary step, the Kostant–Souriau prequantization of the orbit $\mathcal{O}$ [40], [58].

Let $\mathcal{O}$ be an orbit of the coadjoint representation. If $f \in \mathcal{O}$, we have $\mathcal{O} = G \cdot f = G/G(f)$. The stabilizer $G(f)$ of f has Lie algebra $\mathfrak{g}(f)$

$$\mathfrak{g}(f) = \{X \in \mathfrak{g}, f([X, Y]) = 0, \text{ for every } Y \in \mathfrak{g}\}.$$

We can define an alternate non-degenerate 2-form σ_f on the tangent space $\mathfrak{g} \cdot f = \mathfrak{g}/\mathfrak{g}(f)$ to the orbit $\mathcal{O}$ at the point f by the formula $\sigma_f(X \cdot f, Y \cdot f) = f([X, Y])$. This way, we obtain a 2-form σ on $\mathcal{O}$. This form σ gives to $\mathcal{O}$ the structure of a symplectic manifold. If $\dim \mathcal{O} = 2d$, the term $\sigma^d/(d!\,(2\pi)^d)$ of maximal degree of $e^{\sigma/2\pi}$ defines the canonical Liouville measure $dm_{\mathcal{O}}$ on $\mathcal{O}$.

Consider the map $X \to f(X)$ on $\mathfrak{g}(f)$. Clearly $f([X, Y]) = 0$ for $X, Y \in \mathfrak{g}(f)$. We introduce:

$$K(f) = \{\chi, \text{ characters of } G(f) \text{ such that } \chi(\exp X) = e^{i(f, X)} \text{ for } X \in \mathfrak{g}(f)\}.$$

We say that an orbit is integral if $K(f) \neq \emptyset$. (If $G(f)$ is simply connected, $K(f)$ consists of one element.) In particular, f must satisfy the integrality conditions: $(1/2\pi)(f, X) \in \mathbf{Z}$ for all $X \in \mathfrak{g}(f)$ such that $\exp X = e$.

For each character $\chi \in K(f)$, we can construct a line bundle $\mathcal{L}_\chi \to \mathcal{O}$ with $\mathcal{L}_\chi = G \times \mathbf{C}/G(f)$, where $u \in G(f)$ acts on $G \times \mathbf{C}$ by $(g, z) \cdot u = (gu, \chi_f(u)^{-1}z)$. It is not difficult to see that the first Chern class of this line bundle $\mathcal{L}_\chi$ is $\sigma/2\pi$.

Let us say here that the universal formula has some formal analogy with the index formula for the twisted Dirac operator. For example, if G is compact and $\mathcal{O}$ admits a spin structure, the universal formula for $X = 0$ gives us an integral formula for the dimension of the representation $T_{\mathcal{O}}$

$$\dim T_{\mathcal{O}} = \int_{\mathcal{O}} \frac{\sigma^d}{(2\pi)^d d!},$$

which coincides with the index formula ([2]) for the twisted (by $\mathcal{L}_\chi$) Dirac operator D_χ on $\mathcal{O}$. (Recall here that $\mathcal{O}$ was supposed to be of maximal dimension. It follows then from [21] that the tangent bundle to $\mathcal{O}$ is a trivial element of K-theory, thus the term $\hat{\mathcal{A}}$ contributing to the index formula for D_χ is here equal to 1.) Furthermore, Nicole Berline and I [8], have shown that it is indeed possible to give an integral formula for the equivariant index of a connected compact group of transformations of an elliptic complex over a compact manifold, generalizing Kirillov's universal character formula.

In another direction, when $\mathcal{O}$ is not compact but still admits a G-invariant spin structure, the formula of Connes and Moscovici [12] for the L^2-index of D_χ is a precise analogue of Kirillov's formula for $X = 0$. Thus, at least for orbits of maximal dimension with compact stabilizers and spin structures, all these indications would lead us to discover (as Christopher Columbus "discovered" America) the importance of the twisted Dirac operator on orbits to construct the "quantized" representation $T_{\mathcal{O}}$. However, if $\mathcal{O}$ is a general orbit of a group G in $\mathfrak{g}^*$, there is no canonical construction of a representation $T_{\mathcal{O}}$ (even, for some $\mathcal{O}$'s, no construction whatsoever [63]), nor are there sufficiently powerful theorems on unicity of various methods of quantizations. (When G is solvable, the "uniqueness of the quantized orbit" is elucidated in numerous cases. Let us quote the uniqueness theorem of Auslander and Kostant on "independence of positive polarizations" [3] and the uniqueness theorems of Penney [48] and Rosenberg [50] on the realization of $T_{\mathcal{O}}$ in L^2-cohomology spaces.)

3.11. The preceding discussion stressed the importance for the representation theory of G of a subset of orbits satisfying integrality conditions.

Let G be a real algebraic group. By a profound generalization of the result of Kirillov on nilpotent groups, Michel Duflo [17] was able to construct a set X of parameters for $\hat{G}_r$. (When G is semi-simple, this construction is based on Harish-Chandra's work. In case G solvable, a set analogous to X was introduced by Pukanszky [49].) Duflo defines the notion of G-admissible orbits, which is an appropriate modification (see Appendix 1) of the notion of G-integral orbits. In all the examples of the text, the set $\mathfrak{g}_a^*$ of admissible orbits coincide with the set of integral orbits. The set X is given together with a map $d: X \to \mathfrak{g}_a^*/G$ having finite fibers; M. S. Khalgui [34] proved under some very general hypothesis that the universal formula for characters is valid and the map d is indeed the map $T \to \mathcal{O}_T$. We will describe briefly X in Appendix 1. In the examples of this text, we can describe X as follows: Consider the set

$$P = \{(f, \chi); G \cdot f \text{ of maximal dimension}, \chi \in K(f)\}.$$

The group G acts on P. The set P/G is fibered naturally over the set of integral orbits of maximal dimension by the map $d(G \cdot (f, \chi)) = G \cdot f$. We may take for X the set P/G (or a subset with complement of measure 0, as X is defined up to a set of measure 0 for the Plancherel measure $d\mu(T)$).

3.12. Let us describe now for all our examples this fibering and indicate how to prove its compatibility with the universal character formula, i.e. let us relate the character tr T with the Fourier transform of the corresponding orbits $\mathcal{O}_T$.

3.13. Example 4: The Heisenberg Group. The set X is the set of orbits $\mathcal{O}_\lambda$ of maximal dimension where $\mathcal{O}_\lambda = G \cdot \lambda E^*$, $\lambda \neq 0$. Let us consider the character tr T_λ of the representation T_λ (2.12). We will verify that according to the universal character formula (j in this example is identically 1)

$$\operatorname{tr} T_\lambda(\exp X) = \int_{\mathcal{O}_\lambda} e^{i\langle \xi, X\rangle} \frac{\sigma_\lambda}{2\pi},$$

where σ_λ is the canonical 2-form on the orbit $\mathcal{O}_\lambda$.

Indeed, for φ a Schwartz function on $\mathfrak{g}$, we have: (2.12(b))

$$\operatorname{tr}\left(\int T_\lambda(\exp X)\varphi(X)\, dX\right) = \frac{2\pi}{\lambda} \int e^{i\lambda z}\varphi(zE)\, dz,$$

while the second member is

$$\int_{\mathbb{R}^2} \left(\int_{\mathbb{R}^3} e^{i(\lambda E^* + pP^* + qQ^*, xP + yQ + zE)}\varphi(x, y, z)\, dx\, dy\, dz\right) \frac{1}{2\pi\lambda}\, dp\, dq$$

$$= \frac{2\pi}{\lambda} \int e^{i\lambda z}\varphi(0, 0, z)\, dz,$$

by the usual Fourier inversion formula.

3.14. For a simply connected nilpotent group N, the function j is identically 1, and the set X coincides with the set of orbits of maximal dimension. Recall that Kirillov theorem (3.2) gives a description of $\hat{G}$ as $\mathfrak{g}^*/G$. The corresponding character formula

$$\operatorname{tr} T(\exp X) = \int_{\mathcal{O}_T} e^{i\langle \xi, X\rangle} \frac{\sigma^d}{(2\pi)^d d!}$$

holds in fact for every representation T in $\hat{G}$.

More generally, if the representation $T_{\mathcal{O}}$ can be constructed via Mackey induction, the universal character formula holds under certain conditions and can be proven easily ([42]).

Unfortunately, for a general Lie group G, as stressed in 2.10, there is no parametrization of the entire set $\hat{G}$, nor systematic construction of a representation $T_{\mathcal{O}}$ corresponding to an admissible orbit $\mathcal{O}$, if $\mathcal{O}$ is not of maximal dimension. Furthermore, even if $T_{\mathcal{O}}$ is "given" to us, the universal character formula for tr $T_{\mathcal{O}}$ would have to be modified ([33], [8]).

3.15. We now consider the:

EXAMPLE 5: $G = \mathrm{SU}(2)$. We have

$$\mathfrak{g} = \mathfrak{su}(2) = \left\{\begin{pmatrix} ix_3 & -x_1 + ix_2 \\ x_1 + ix_2 & -ix_3 \end{pmatrix}, x_i \in \mathbf{R}\right\}.$$

We identify $\mathfrak{g}$ with $\mathfrak{g}^*$ via the G-invariant bilinear form $(X, Y) \to -\frac{1}{2}\operatorname{Tr}(XY)$. Recall that the function $X \to \det X$ is invariant by the adjoint action of G on $\mathfrak{g}$. Thus the orbits of G in $\mathfrak{g}^*$ are the spheres $x_1^2 + x_2^2 + x_3^2 = r^2$.

If

$$f = \begin{pmatrix} i\lambda & 0 \\ 0 & -i\lambda \end{pmatrix} \in \mathfrak{g}^* \qquad (\lambda \neq 0),$$

we have

$$\mathfrak{g}(f) = \left\{\begin{pmatrix} i\theta & 0 \\ 0 & -i\theta \end{pmatrix}; \theta \in \mathbf{R}\right\},$$

$$G(f) = \left\{\begin{pmatrix} e^{i\theta} & 0 \\ 0 & e^{-i\theta} \end{pmatrix}; \theta \in \mathbf{R}\right\}.$$

The form f is integral if $\chi_f\begin{pmatrix} e^{i\theta} & 0 \\ 0 & e^{-i\theta} \end{pmatrix} = e^{i\lambda\theta}$ is well defined on $G(f)$, i.e. if $\lambda \in \mathbf{Z} - \{0\}$. As $G(f)$ is connected, $K(f)$ is either empty or, if f is integral, it is the set with the one element χ_f.

Thus the set $\mathfrak{g}_G^*$ of integral orbits of maximal dimension coincides with all spheres S_n with positive integral radius. We may picture the set $\mathfrak{g}_G^*$ and X by:

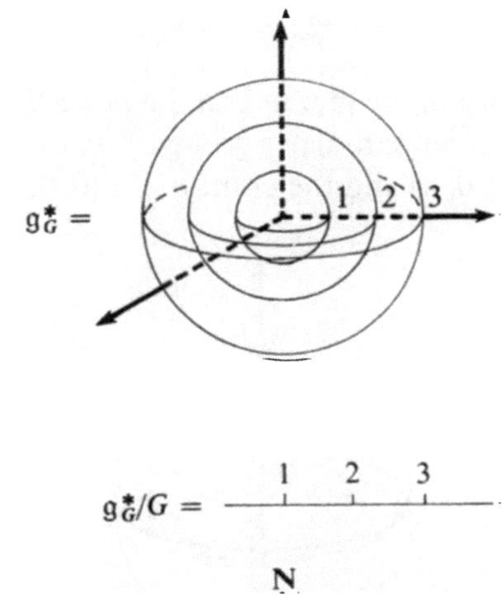

Figure 8

As we compare (2.13), Figure 5 with Figure 8, we recovered our description of $\hat{G} = \{T_n; n \in N\}$.

Let us describe the relation of the character of T_n with the Fourier transform of the measure on the sphere S_n. It is easy to see that the function $j(X)$ has an analytic square root on all of $\mathfrak{g}$ and that

$$j^{1/2}\begin{pmatrix}\theta & 0\\ 0 & -\theta\end{pmatrix} = \frac{e^{i\theta} - e^{-i\theta}}{2i\theta}.$$

The universal character formula asserts that

$$\operatorname{tr} T_n\begin{pmatrix}e^{i\theta} & 0\\ 0 & e^{-i\theta}\end{pmatrix} j^{1/2}\begin{pmatrix}\theta & 0\\ 0 & -\theta\end{pmatrix} = \int_{x_1^2+x_2^2+x_3^2=n^2} e^{ix_3\theta}\,\frac{\sigma}{2\pi},$$

i.e.

$$\frac{e^{in\theta} - e^{-in\theta}}{e^{i\theta} - e^{-i\theta}}\,\frac{e^{i\theta} - e^{-i\theta}}{2i\theta} = \int_{x_1^2+x_2^2+x_3^2=n^2} e^{ix_3\theta}\,\frac{\sigma}{2\pi}.$$

Thus we have to prove:

$$\frac{e^{in\theta}}{2i\theta} - \frac{e^{-in\theta}}{2i\theta} = \int_{x_1^2+x_2^2+x_3^2=n^2} e^{ix_3\theta}\,\frac{\sigma}{2\pi},$$

for $\sigma/2\pi$ the canonical Liouville measure on the orbit S_n of SU(2).

It is immediate to verify this formula using spherical coordinates (see Appendix 2).

3.16. For a compact Lie group G, the universal character formula is equivalent to a well-known formula of Harish-Chandra, established long before. Developing an earlier idea of R. Bott [10], Nicole Berline and I [7] have given a simpler proof of a more general formula of Duistermaat–Heckmann for torus actions on symplectic manifolds [22].

We may explain the idea of the method on the preceding Example 5 as follows: Consider the action of the one-parameter group $e^{i\theta}$ on the orbit S_n. It has two fixed points, the point $p^+ = (0, 0, n)$ and the point $p^- = (0, 0, -n)$.

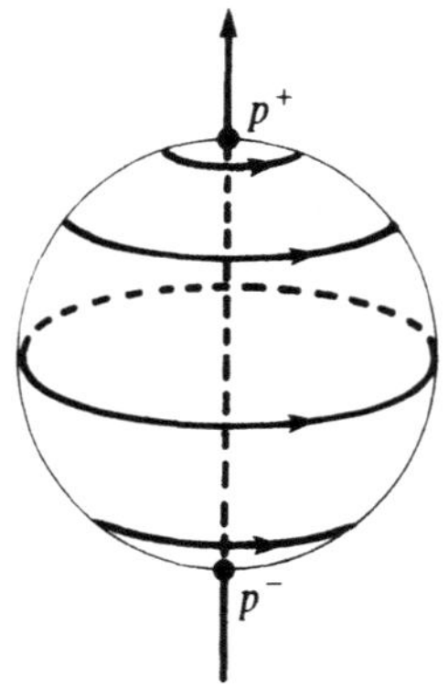

Figure 9

It is easy to see that the form to be integrated is exact, except at these two points and that the two terms $e^{in\theta}/2i\theta$ and $e^{-in\theta}/2i\theta$ comes from a calculus of residues at these two points.

3.17. Let us finally consider our

EXAMPLE 6: $G = \mathrm{SL}(2, \mathbf{R})$. We have

$$\mathfrak{g} = \mathfrak{sl}(2, \mathbf{R})$$
$$= \left\{ \begin{pmatrix} x_1 & x_2 + x_3 \\ x_2 - x_3 & -x_1 \end{pmatrix}; x_i \in \mathbf{R} \right\}.$$

We identify $\mathfrak{g}$ with $\mathfrak{g}^*$ via the G-invariant bilinear form $(X, Y) \to \frac{1}{2}\operatorname{tr}(XY)$. Recall that the function $\det X = x_3^2 - (x_1^2 + x_2^2)$ is invariant by the action of G on $\mathfrak{g}^*$. From this, it follows that the orbits of G in $\mathfrak{g}^*$ are:

(1) (a) The upper sheet $x_3 \geq 0$ of the two-sheeted hyperboloid

$$x_3^2 - (x_1^2 + x_2^2) = \lambda^2, \qquad \lambda \neq 0.$$

A typical element of this orbit is

$$f = \begin{pmatrix} 0 & \lambda \\ -\lambda & 0 \end{pmatrix}, \qquad \lambda > 0.$$

(b) The lower sheet $x_3 \leq 0$ of the two-sheeted hyperboloid

$$x_3^2 - (x_1^2 + x_2^2) = \lambda^2, \qquad \lambda \neq 0.$$

A typical element of this orbit is

$$f = \begin{pmatrix} 0 & \lambda \\ -\lambda & 0 \end{pmatrix}, \qquad \lambda < 0.$$

We shall denote by $\mathcal{O}_\lambda^d$ the orbit of the element $f = \begin{pmatrix} 0 & \lambda \\ -\lambda & 0 \end{pmatrix}$.

(2) The one-sheeted hyperboloid $x_3^2 - (x_1^2 + x_2^2) = -s^2 (s \neq 0)$.

A typical element of this orbit is $f = \begin{pmatrix} s & 0 \\ 0 & s \end{pmatrix}$. We denote by $\mathcal{O}_s^p$ the orbit of the element $\begin{pmatrix} s & 0 \\ 0 & -s \end{pmatrix}$.

(3) The point $\{0\}$ and the two connected components of the light cone

$$(x_3^2 - x_1^2 - x_2^2 = 0, x \neq 0).$$

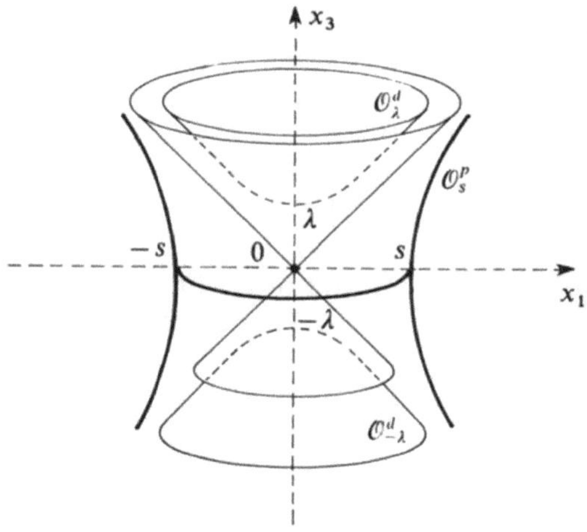

Figure 10

We determine now the set of integral orbits:

(1) For

$$f = \begin{pmatrix} 0 & \lambda \\ -\lambda & 0 \end{pmatrix} \in \mathfrak{g}^* \qquad (\lambda \neq 0),$$

we have:

$$\mathfrak{g}(f) = \left\{ \begin{pmatrix} 0 & \theta \\ -\theta & 0 \end{pmatrix}; \theta \in \mathbf{R} \right\},$$

$$G(f) = \left\{ \begin{pmatrix} \cos\theta & \sin\theta \\ -\sin\theta & \cos\theta \end{pmatrix}; \theta \in \mathbf{R} \right\}.$$

The form f is integral if

$$\chi_f\begin{pmatrix} \cos\theta & \sin\theta \\ -\sin\theta & \cos\theta \end{pmatrix} = e^{i\lambda\theta}$$

is well defined on $G(f)$, i.e. if $\lambda \in \mathbf{Z} - \{0\}$. In this case ($\lambda$ integer), as $G(f)$ is connected, $K(f)$ consists of one element, the character χ_f.

(2) For

$$f = \begin{pmatrix} s & 0 \\ 0 & -s \end{pmatrix} \in \mathfrak{g}^*,$$

$$\mathfrak{g}(f) = \left\{\begin{pmatrix} t & 0 \\ 0 & -t \end{pmatrix}; t \in \mathbf{R}\right\},$$

$$G(f) = \left\{\begin{pmatrix} a & 0 \\ 0 & a^{-1} \end{pmatrix}; a \in \mathbf{R} - \{0\}\right\}.$$

Every f of this form is integral, and for any s, the set $K(f)$ consists of two elements, the characters χ_s^+ and χ_s^- with

$$\chi_s^+\begin{pmatrix} \varepsilon e^t & 0 \\ 0 & \varepsilon e^{-t} \end{pmatrix} = e^{ist}, \qquad \varepsilon = \pm 1,$$

$$\chi_s^-\begin{pmatrix} \varepsilon e^t & 0 \\ 0 & \varepsilon e^{-t} \end{pmatrix} = \varepsilon e^{ist}.$$

We denote by $\mathfrak{g}_G^*$ the set of integral orbits of maximal dimension of the sets (1) and (2) (i.e. we omit the light cone). Thus $\mathfrak{g}_G^*$ consists of the orbits:

$$\mathcal{O}_n^d = G \cdot \begin{pmatrix} 0 & n \\ -n & 0 \end{pmatrix}, \qquad n > 0,$$

$$\mathcal{O}_n^d = G \cdot \begin{pmatrix} 0 & n \\ -n & 0 \end{pmatrix}, \qquad n < 0,$$

$$\mathcal{O}_s^p = G \cdot \begin{pmatrix} s & 0 \\ 0 & -s \end{pmatrix}, \qquad s \neq 0.$$

Thus we obtain Figure 11 for $\mathfrak{g}_G^*$ and $\mathfrak{g}_G^*/G$.

The set $X = \hat{G}_r$ is given by X^{irr}/G with $X^{\mathrm{irr}} = \{(f, \chi), f \in \mathfrak{g}_G^*, \chi \in K(f)\}$. Thus X is fibered over $\mathfrak{g}_G^*/G$ with fibers consisting of the one point T_n over the orbit $\mathcal{O}_n^d$ (d for discrete) and of the two points $T_s^{\pm}$ over the orbit $\mathcal{O}_s^p$ (p for principal). The reader may then compare Figure 10 with Figure 5 (2.14) to visualize the fibering.

The universal character formula for the representation T_n associated to the orbit $\mathcal{O}_n$ is equivalent to the equality of generalized functions ($n > 0$):

$$\text{“}\frac{e^{in\theta}}{2i\theta} = -\int_{\substack{x_3^2-(x_1^2+x_2^2)=n^2 \\ x_3 \geq 0}} e^{ix_3\theta}\,\frac{\sigma}{2\pi}\text{,”}$$

when $\sigma/2\pi$ is the canonical 2-form on $\mathcal{O}_n$.

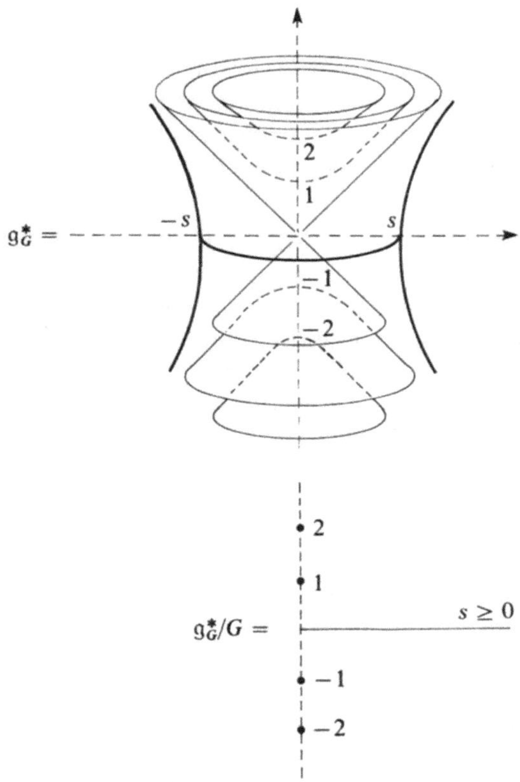

Figure 11

3.18. The validity of this formula or more generally the validity of the universal character formula for a semi-simple Lie group was established by W. Rossmann [51]: This fundamental result gave legitimacy to the claims of universality for the orbit method.

It is possible to generalize the argument sketched in 3.16 to prove Rossmann's formula [7]:

Let us remark, in the case of SL(2, **R**), that the action of the one-parameter group $\begin{pmatrix} \cos\theta & \sin\theta \\ -\sin\theta & \cos\theta \end{pmatrix}$ on $\mathcal{O}_n$ has only one fixed point, the point $p_0 = (0, 0, n)$.

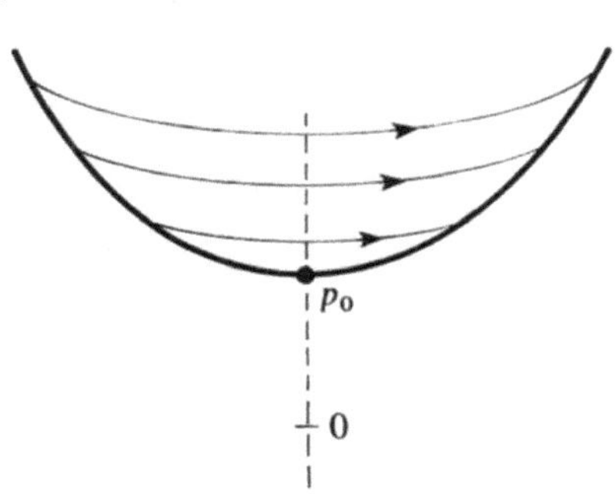

Figure 12

As in the case of SU(2), the form to be integrated is exact except at this point. Thus there is only one residue to be calculated which leads to the term $-e^{in\theta}/2i\theta$.

Similarly, the character formula $\Theta_\lambda(g)$ for the representation T_λ of the discrete series has a simple geometric interpretation as a fixed point formula, for the elements g belonging to the elliptic set. However, the general formulae of Rebecca Herb for arbitrary regular elements g of G are not yet reducible to a simple geometric interpretation.

3.19. We have related in Examples 1–6 the set $\hat{G}_r$ with the geometric set X, and the distribution tr T with Fourier transforms of canonical measures on orbits. Thus, due to the work of Michel Duflo, we see that the set $\hat{G}_r$ is described adequately. It is still, however, an open question to determine explicitly for a general Lie group G the corresponding Plancherel measure on X.[1] When G is a simply connected nilpotent Lie group, the set $\hat{G}$ is merely the set of orbits $\mathfrak{g}^*/G$. Each orbit $\mathcal{O}$ has a canonical measure $dm_{\mathcal{O}}$. Let dg be a Haar measure on G, dX the corresponding measure on $\mathfrak{g}$, df the dual Haar measure on $\mathfrak{g}^*$. The usual Fourier inversion formula on the vector space

$$\varphi(0) = \int_{\mathfrak{g}^*} \hat{\varphi}(f)\, df,$$

yield immediately to the Plancherel formula:

$$\tilde{\varphi}(e) = \varphi(0) = \int_{\mathfrak{g}^*/G} \left(\int_{\mathcal{O}} \hat{\varphi}\, dm_{\mathcal{O}} \right) dp(\mathcal{O}) = \int_{\mathfrak{g}^*/G} \operatorname{tr}(T_{\mathcal{O}}, \tilde{\varphi})\, dp(\mathcal{O}),$$

with $\tilde{\varphi}$ the C^∞-function on G such that $\varphi = \tilde{\varphi} \cdot \exp$, and $dp = df/dm_{\mathcal{O}}$ the quotient measure of df by the canonical measures $dm_{\mathcal{O}}$. A similar formula holds for a solvable simply-connected type I unimodular Lie group G. (Pukanszky [49], Charbonnel [11]).

In general, as we have seen, orbits $\mathcal{O}_T$ corresponding to representations must satisfy some non-empty integrality conditions and the corresponding set $\mathfrak{g}^*_G = \{\bigcup \mathcal{O}_T, T \in \hat{G}_r\}$ is not dense in $\mathfrak{g}^*$.

The case of a torus $T = V/\Gamma$ is instructive. In this case, $\mathfrak{g} = V$ and $\mathfrak{g}^* = V^*$, while the set of orbits corresponding to representations of T is the discrete subset $\Gamma^* \subset V^*$. Let dx be the Euclidean measure on V giving measure 1 to the fundamental parallelepiped on Γ. The Poisson summation formula is:

$$\sum_{\gamma \in \Gamma} \varphi(\gamma) = \sum_{\gamma^* \in \Gamma^*} \varphi(\gamma^*),$$

that we may rewrite as:

$$\sum_{\substack{\gamma \in \mathfrak{g} \\ \exp \gamma = e}} \varphi(\gamma) = \sum_{\gamma^* \in G \subset \mathfrak{g}^*} \varphi(\gamma^*).$$

Figure 13

[1] The Plancherel measure is now determined (M. Duflo, special year in Lie groups representations, Maryland, 1983).

3.20. Consider for a general Lie group G the maps $\exp: \mathfrak{g} \to G$ and $d: \hat{G}_r \to \mathfrak{g}^*/G$, (we write $d(T) = \mathcal{O}_T$) and the subsets

$$\mathfrak{g}_G = \{X \in \mathfrak{g};\, \exp X = e\} \quad \text{of } \mathfrak{g},$$

$$\mathfrak{g}^*_{\hat{G}} = \bigcup \mathcal{O}_T, \quad T \in \hat{G}_r \qquad \text{of } \mathfrak{g}^*.$$

Recall that $\mathfrak{g}^*_{\hat{G}}$ is determined in purely geometric terms by some integrality conditions (see Appendix 1). In the examples given in this text, $\mathfrak{g}^*_{\hat{G}}$ may be taken as the set of integral orbits of maximum dimension. I conjectured [61] that a similar "Poisson formula" relates the sets $\mathfrak{g}_G$ and $\mathfrak{g}^*_{\hat{G}}$. Let us formulate this conjecture in these terms (see Appendix 1 for a more precise formulation): "There exists a G-invariant positive generalized function v_G on $\mathfrak{g}^*$ of support $\mathfrak{g}^*_{\hat{G}}$ whose Fourier transform is a distribution n_G, such that the support of n_G is contained in $\mathfrak{g}_G$ and n_G coincides with $\delta(0)$ near the origin."

In the case of the torus T, v_G is the δ-function of the lattice Γ^* and n_G is the δ-distribution of the lattice Γ.

Of course if G is a simply connected nilpotent Lie group, we can take $n_G = \delta(0)$ and $v_G = 1$. I proved that this conjecture holds for G semi-simple linear [62], and Duflo proved that this conjecture holds for G any complex Lie group [19].

3.21. Let us now examine briefly in Examples 5 and 6 the content of this conjecture. As it will be clear on the examples, the form of the generalized function v_G is prescribed by the Plancherel formula of G.

3.22. Example 5: $G = \mathrm{SU}(2)$. Recall that the orbits of G in $\mathfrak{g}(\simeq \mathfrak{g}^*)$ are the spheres $S_r = \{x_1^2 + x_2^2 + x_3^2 = r^2\}$ $(r \geq 0)$.

As

$$\exp\begin{pmatrix} i\theta & 0 \\ 0 & -i\theta \end{pmatrix} = \begin{pmatrix} e^{i\theta} & 0 \\ 0 & e^{-i\theta} \end{pmatrix}$$

the set $\mathfrak{g}_G$ consists of all orbits S_r of radius $r = (0, 2\pi, 4\pi, 6\pi, \ldots)$.

Recall that $\mathfrak{g}^*_{\hat{G}}$ consists of all orbits S_r of non-zero integral radius $(1, 2, 3, \ldots)$.

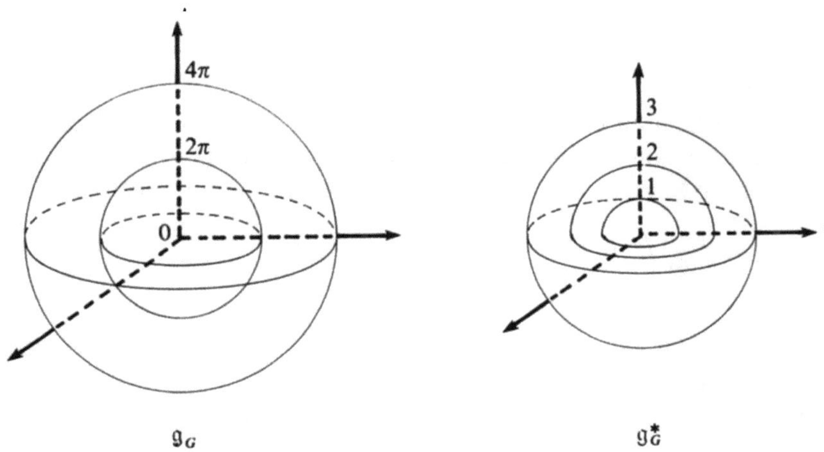

Figure 14

Let, for $a \in \mathbf{R}$, Θ_a be the G-invariant function on $\mathfrak{g}^*$ such that

$$\Theta_a\begin{pmatrix} i\lambda & 0 \\ 0 & -i\lambda \end{pmatrix} = \tfrac{1}{2}(e^{i\lambda a} + e^{-i\lambda a}).$$

Let v_G be the G-invariant generalized function on $\mathfrak{g}^*$ given by $v = \sum_{a \in 2\pi\mathbf{Z}} \Theta_a$, i.e.:

$$v_G\begin{pmatrix} i\lambda & 0 \\ 0 & -i\lambda \end{pmatrix} = \sum_n e^{2i\pi n\lambda}.$$

It is not difficult to prove (see Appendix 2) that for $df = (1/4\pi)\, d\xi_1\, d\xi_2\, d\xi_3$ on $\mathfrak{g} \simeq \mathfrak{g}^*$

$$\int_{\mathfrak{g}^*} v_G(f)\varphi(f)\, df = \sum_{n \in \mathbf{N}} n\left(\int_{S_n} \varphi \frac{\sigma_n}{2\pi}\right),$$

where $\sigma_n/2\pi$ is the canonical Liouville measure on S_n. Thus $v_G(f)df$ is a positive measure supported on $\mathfrak{g}_G^*$ (this measure is clearly derived from the Plancherel formula on $\hat{G}$). The proof of this equality follows from the usual Poisson summation formula:

$$\sum_n e^{2i\pi nx}\, dx = \sum \delta(n).$$

In Appendix 2, we will compute the Fourier transform n_G of v_G and show that n_G satisfies the required properties of the conjecture, in particular that n_G is a distribution of support $\mathfrak{g}_G$.

3.23. Example 6: $G = \mathrm{SL}(2, \mathbf{R})$. Denote by $\mathcal{O}_\lambda^d$ the orbit of the element $\begin{pmatrix} 0 & \lambda \\ -\lambda & 0 \end{pmatrix}$ and by $\mathcal{O}_s^p$ the orbit of the element $\begin{pmatrix} s & 0 \\ 0 & -s \end{pmatrix}$. As

$$\exp\begin{pmatrix} 0 & \theta \\ -\theta & 0 \end{pmatrix} = \begin{pmatrix} \cos\theta & \sin\theta \\ -\sin\theta & \cos\theta \end{pmatrix},$$

the set $\mathfrak{g}_G$ consists of all orbits $\mathcal{O}_a^d$, for $a \in 2\pi\mathbf{Z}$. Recall that the set $\mathfrak{g}_G^*$ consists of the orbits $\mathcal{O}_n^d$, for n non-zero integer and of all the orbits $\mathcal{O}_s^p$ $(s \neq 0)$, see Figure 15.

Define, for $a \in \mathbf{R}$, Θ_a as being the G-invariant function on $\mathfrak{g}^*$ such that

$$\Theta_a\begin{pmatrix} 0 & \lambda \\ -\lambda & 0 \end{pmatrix} = e^{i\lambda a}, \qquad \Theta_a\begin{pmatrix} s & 0 \\ 0 & -s \end{pmatrix} = e^{-|as|}.$$

(Θ_a is defined by this relation except on the light cone, but, as Θ_a is bounded, Θ_a defines unambiguously a generalized function on $\mathfrak{g}^*$.)

Define $v_G = \sum_{a \in 2\pi\mathbf{Z}} \Theta_a$. It is not difficult to prove (see Appendix 3) that for $df = (1/4\pi)\, dx_1\, dx_2\, dx_3$ on $\mathfrak{g} \simeq \mathfrak{g}^*$

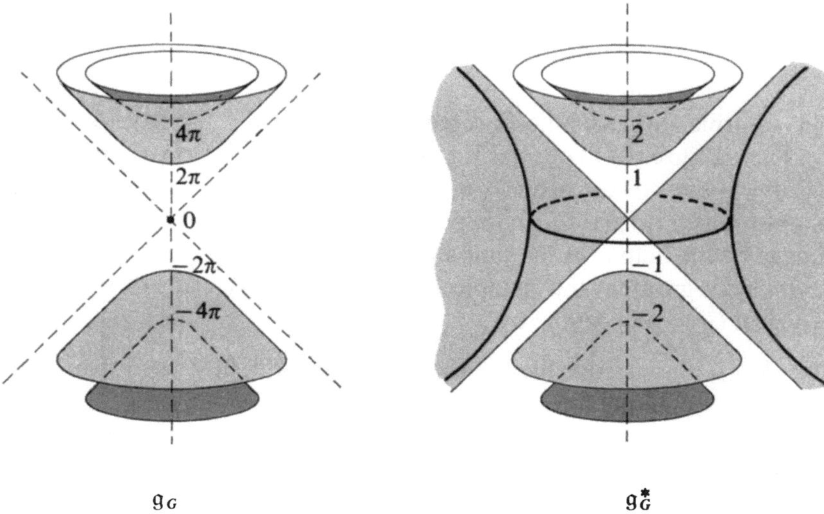

Figure 15

3.24 $$\int_{\mathfrak{g}^*} v_G(f)\varphi(f)\,df = \sum_{n\in\mathbf{Z}} |n| \left(\int_{\mathcal{O}_n^d} \varphi\,\frac{\sigma}{2\pi}\right) + \int_0^\infty s\coth \pi s\left(\int_{\mathcal{O}_s^p} \varphi\,\frac{\sigma}{2\pi}\right) ds.$$

Then $v_G(f)\,df$ is a positive measure supported on $\mathfrak{g}_G^*$. (This measure is clearly derived from the image of the Plancherel measure on $\hat{G}_r$ by the map $\hat{G}_r \to \mathfrak{g}_G^*/G$. Recall that the fiber of this map above $\mathcal{O}_s$ consists of two points T_s^+ and T_s^- with respective weights in the Plancherel measure $\frac{1}{2}s$ than $(\pi s/2)$ and $\frac{1}{2}s$ coth$(\pi s/2)$ and that $\frac{1}{2}$ than $(\pi s/2) + \frac{1}{2}$ coth$(\pi s/2) = \coth \pi s$.)

The proof of the formula 3.24 relies on the two identities:

$$\sum_{n\in\mathbf{Z}} e^{2i\pi n\lambda}\,d\lambda = \sum \delta(n),$$

$$\sum_{n\in\mathbf{Z}} e^{-|2\pi n s|} = \coth \pi s \quad \text{for } s \neq 0.$$

Let us remark that the left-hand side of these two identities involves the formulas for the characters of the discrete series T_n, respectively, on the element $\begin{pmatrix} 0 & \lambda \\ -\lambda & 0 \end{pmatrix}$ and on the element $\begin{pmatrix} s & 0 \\ 0 & -s \end{pmatrix}$. The proof of the conjecture for general linear semi-simple Lie groups is based on similar equalities between the discrete series constants and the Plancherel functions ([62]).

The proof of this conjecture for non-linear semi-simple Lie groups would lead to some better understanding of the relation between character formulae, integral orbits, and the Plancherel measure on G.

I hope many mathematicians, women and men, will continue to work on the topics touched upon here.

APPENDIX 1

Duflo's Parametrization of $\hat{G}_r$

Let $\mathcal{O}$ be an orbit of the co-adjoint representation. If $f \in \mathcal{O}$, we have $\mathcal{O} = G \cdot f = G/G(f)$. The stabilizer $G(f)$ of f has Lie algebra $\mathfrak{g}(f)$. The 2-form σ_f on the tangent space $\mathfrak{g} \cdot f = \mathfrak{g}/\mathfrak{g}(f)$ to the orbit $\mathcal{O}$ at the point f is invariant by $\tilde{G}(f)$. Thus we obtain a morphism i_f from $G(f)$ to $\mathrm{Sp}(\mathfrak{g}/\mathfrak{g}(f))$. Consider the 2-fold cover $\tilde{G}(f) \to G(f)$ image by the map i_f of the canonical 2-fold cover $\mathrm{Mp}(\mathfrak{g}/\mathfrak{g}(f)) \to \mathrm{Sp}(\mathfrak{g}/\mathfrak{g}(f))$ of the symplectic group by the metaplectic group. Let (ε, e) be the reciprocal image of $e \in G(f)$ in $\tilde{G}(f)$. We denote by

$$X(f) = \{\tau, \text{ irreducible representations of } G(f) \text{ in } V_\tau \text{ such that}$$
$$(1)\ \tau(\exp X) = e^{i\langle f, X\rangle}\,\mathrm{Id}_{V_\tau} \text{ for } X \in \mathfrak{g}(f);$$
$$(2)\ \tau(\varepsilon) = -\mathrm{Id}_{V_\tau}\}.$$

An element $f \in \mathfrak{g}^*$ is called admissible if $X(f)$ is not empty. An element $f \in \mathfrak{g}^*$ is called regular if $\mathcal{O} = G \cdot f$ is of maximal dimension.

Let G be a real algebraic group. The set $X(f)$ consists of a finite number of finite dimensional representations of $\tilde{G}(f)$. Let f be regular, then the connected component $G(f)^0$ of $G(f)$ is commutative and is the direct product of its semi-simple part $S(f)$ with its unipotent part $U(f)$. Call f strongly regular, if $S(f)$ is of maximal dimension (among the subgroups $S(f)$, for f regular). The conjugacy classes of the subgroups $S(f)$, f strongly regular, are in finite number. (If G is a complex group, the subgroups $S(f)$, f strongly regular, are in the same conjugacy class. This is not the case of the subgroups $U(f)$ [39]). If G is a semi-simple group, the subgroup $S(f)$, for f strongly regular, is a Cartan subgroup of G.

Denote by

$$\mathfrak{g}^*_G = \{f \in \mathfrak{g}^*, f \text{ admissible}, f \text{ strongly regular}\},$$

$$\mathfrak{g}_G = \{X \in \mathfrak{g};\ \exp X = e\}.$$

Define:

$$X^{\mathrm{irr}} = \{(f, \tau);\ f \in \mathfrak{g}^*_G,\ \tau \in X(f)\}.$$

The group G acts on X^{irr}. Define $X = X^{\mathrm{irr}}/G$. Duflo constructed an application $(f, \tau) \mapsto T_{f,\tau}$ of X^{irr} in $\hat{G}$, which induces an injective application d of X into $\hat{G}$. Finally, under some hypothesis which are probably automatically satisfied, Duflo proved that the Plancherel measure $d\mu(T)$ is concentrated on $d(X)$. Thus we may consider that the problem of determining $\hat{G}_r$ is entirely solved by these results.

Let us relate this parametrization X of $\hat{G}_r$ with the Kirillov character formula. Let σ be the canonical 2-form on the orbit $\mathcal{O}$ of dimension $2d$. Consider $e^{\sigma/2\pi}$ and its term $(1/d!)(\sigma^d/(2\pi)^d)$ of maximal degree. Let G be unimodular, $f \in \mathfrak{g}^*_G$ such that $\mathcal{O} = G \cdot f$ is closed, $\tau \in X(f)$, then M. S. Khalgui proved that indeed:

$$\mathrm{tr}\, T_{f,\tau}(\exp X) j(X)^{1/2} = \int_{\mathcal{O}} (\dim \tau) e^{i\langle \xi, X\rangle} \frac{1}{d!} \frac{\sigma^d}{(2\pi)^d},$$

as an equality of generalized functions in a neighborhood of O. [34].

(For $\mathfrak{g}$ semi-simple, a corner-stone case, this was due to Rossman [51]. For $\mathfrak{g}$

solvable, it was proven by Duflo [9].) Thus the map $(f, \tau) \mapsto f$ of X^{irr} into $\mathfrak{g}_G^*$ induces the Kirillov map $T \mapsto \mathcal{O}_T$ from $\hat{G}_r$ to $\mathfrak{g}_G^*/G$.

Let us now formulate my conjecture on the Poisson–Plancherel formula. Let dg be a Haar measure on G. Let dX be the corresponding Euclidean measure on $\mathfrak{g}$: i.e. for φ supported in a small neighborhood of 0 (where the exponential map is a diffeomorphism) $\int_G \tilde{\varphi}(g)\, dg = \int_{\mathfrak{g}} \varphi(X) j(X)\, dX$, with $\tilde{\varphi}$ the function on G such that $\varphi = \tilde{\varphi} \cdot \exp$. Let df be the dual measure to dX on $\mathfrak{g}^*$, i.e. df is such that

$$\varphi(0) = \int_{\mathfrak{g}^*} \left(\int_{\mathfrak{g}} e^{i\langle f, X\rangle} \varphi(X)\, dX \right) df = \int_{\mathfrak{g}^*} \hat{\varphi}(f)\, df.$$

Let $d\mu(f, \tau)$ be the Plancherel measure on X. Denote by $\tilde{\mu}$ the measure on $\mathfrak{g}_G^*/G$ image of the measure $(\dim \tau)\, d\mu(f, \tau)$.

Let $\hat{\mu}$ be the positive measure on $\mathfrak{g}_G^*$ such that:

$$\int_{\mathfrak{g}_G^*} \alpha(f)\, d\hat{\mu}(f) = \int_{\mathfrak{g}_G^*/G} \left(\int_{\mathcal{O}} \alpha(f)\, dm_{\mathcal{O}}(f) \right) d\tilde{\mu}(\mathcal{O}).$$

Let $\tilde{\varphi}$ be a function on G supported in a small neighborhood of e and φ the function on $\mathfrak{g}$ such that $\varphi = \tilde{\varphi} \circ \exp$. We have

$$\begin{aligned}
(\operatorname{tr} T_{f,\tau}, \tilde{\varphi}) &= \operatorname{tr} \int_G T_{f,\tau}(g) \tilde{\varphi}(g)\, dg \\
&= \operatorname{tr} \int_{\mathfrak{g}} T_{f,\tau}(\exp X) \tilde{\varphi}(\exp X) j(X)^{1/2} j(X)^{1/2}\, dX \\
&= (\dim \tau) \int_{\mathcal{O} = G \cdot f} (\varphi j^{1/2})^{\wedge}(l)\, dm_{\mathcal{O}}(l).
\end{aligned}$$

Thus, from the Plancherel formula on G, we obtain

$$\begin{aligned}
\varphi(0) = \tilde{\varphi}(e) &= \int_X \operatorname{tr}(T_{f,\tau}, \tilde{\varphi})\, d\mu(f, \tau) \\
&= \int_X (\dim \tau) \int_{\mathcal{O} = G \cdot f} (\varphi j^{1/2})^{\wedge}(l)\, dm_{\mathcal{O}}(l)\, d\mu(f, \tau) \\
&= \int_{\mathfrak{g}_G^*/G} \left(\int_{\mathcal{O}} (\varphi j^{1/2})^{\wedge}(l)\, dm_{\mathcal{O}}(l) \right) d\tilde{\mu}(\mathcal{O}) \\
&= \int_{\mathfrak{g}_G^*} (\varphi j^{1/2})^{\wedge}(f)\, d\hat{\mu}(f).
\end{aligned}$$

Let v_G be the positive generalized function on $\mathfrak{g}^*$ concentrated on $\mathfrak{g}_G^*$ such that $d\hat{\mu}(f) = v_G(f)\, df$. We then see from this formula that, if n_G is the Fourier transform of v_G, then n_G is a distribution on $\mathfrak{g}$ which coincides with the Dirac measure at 0.

I formulated the following conjecture (in somewhat more timid terms; I am indebted to Michel Duflo [20] for the present reformulation):

Let G be a Lie group with Lie algebra $\mathfrak{g}$. There exists a tempered distribution n_G on $\mathfrak{g}$ having the following properties:

(1) *n_G is G-invariant;*

(2) *n_G coincides with the Dirac measure in a neighborhood of* 0; *the support of n_G is contained in $\mathfrak{g}_G$;*
(3) *n_G is of positive type and its Fourier transform is a generalized positive function v_G concentrated on $\mathfrak{g}_G^*$;*
(4) *the function $j(X)^{1/2}$ admits an analytic square root in a neighborhood of the support of n_G and $j(X)^{1/2}n_G$ is a measure.*

(It is clear that it is also expected, from the preceding discussion, when G is a type I unimodular group and X the Duflo set of parameters for $\hat{G}_r$ that $v_G\, df = d\hat{\mu}(f)$.)

APPENDIX 2

Complements on SU(2)

1. Normalization of Measures

Let

$$G = \left\{\begin{pmatrix} \alpha & -\bar{\beta} \\ \beta & \bar{\alpha} \end{pmatrix}; |\alpha|^2 + |\beta|^2 = 1\right\} = \mathrm{SU}(2),$$

$$\mathfrak{g} = \left\{\begin{pmatrix} ix_3 & -x_1 + ix_2 \\ x_1 + ix_2 & -ix_3 \end{pmatrix}; x_i \in \mathbf{R}\right\}$$

the Lie algebra of G with corresponding basis J_1, J_2, J_3.

The map $g \mapsto (\alpha, \beta)$ identifies G with

$$S_3 = \{(y_1 + iy_2, y_3 + iy_4); y_1^2 + y_2^2 + y_3^2 + y_4^2 = 1\}.$$

Let dg be the Haar measure on G giving total mass 1 to G. If μ is the surface measure on S_3, then $dg = \mu/2\pi^2$. The Haar measure is left and right invariant.

We consider the corresponding Euclidean measure dX on $\mathfrak{g}$ (i.e. $dg = j(X)\, dX$). We have:

$$dX = \frac{dx_1\, dx_2\, dx_3}{2\pi^2}.$$

2. Weyl Integration Formula

Proposition. *Let f be a continuous function on G, then:*

$$\int_G f(g)\, dg = \frac{1}{4\pi}\int_0^{2\pi} |e^{i\theta} - e^{-i\theta}|^2 \int_G f\left(g\begin{pmatrix} e^{i\theta} & 0 \\ 0 & e^{-i\theta} \end{pmatrix}g^{-1}\right) dg\, d\theta.$$

Proof. Recall that an element g of G can be conjugated to an element

$$\left[\begin{pmatrix} e^{i\theta} & 0 \\ 0 & e^{-i\theta} \end{pmatrix}; \qquad 0 \leq \theta \leq \pi.\right.$$

$\begin{pmatrix} e^{i\theta} & 0 \\ 0 & e^{-i\theta} \end{pmatrix}$ is itself conjugated to $\begin{pmatrix} e^{-i\theta} & 0 \\ 0 & e^{i\theta} \end{pmatrix}$ by $w = \left.\begin{pmatrix} 0 & 1 \\ -1 & 0 \end{pmatrix}\right]$.

Let $T = \left\{\begin{pmatrix} e^{i\theta} & 0 \\ 0 & e^{-i\theta} \end{pmatrix}; \theta \in \mathbf{R}\right\}$ and consider $P = G/T$. We denote by $\dot{g}$ the image of g on G/T. There exists a unique G-invariant measure $d\dot{g}$ on G/T such that $\int_{G/T} f(\dot{g})\, d\dot{g} = \int_G f(\dot{g})\, d\dot{g}$. Identify the tangent space at $\dot{e}$ with $\mathbf{R}J_1 \oplus \mathbf{R}J_2$. The corresponding volume form ω at $\dot{e}$ is such that $\omega(J_1 \Lambda J_2) = 1/\pi$. The map $c(g, \theta) = g\begin{pmatrix} e^{i\theta} & 0 \\ 0 & e^{-i\theta} \end{pmatrix} g^{-1}$ is a diffeomorphism of $P \times]0, \pi[$ with the subset $G'_r = \{g \in G;\ g \neq (1, -1)\}$ of G. Thus there exists a measure $\mu(x, \theta)$ on $P \times]0, \pi[$, such that

$$\int f(g)\, dg = \int_{P \times]0, \pi[} f(c(x, \theta))\mu(x, \theta).$$

Using the Ad G-invariance of dg, we see that $\mu(x, \theta) = J(\theta)\, d\theta\, dg$. To compute $J(\theta)$, we need then to compute the Jacobian of c at the point $(\dot{e}, \theta)$. In coordinates (y_1, y_2, y_3, y_4) for $S_3 \simeq G$, it is immediate to see that

$$c_*(J_1) = \frac{d}{d\varepsilon}\begin{pmatrix} 1 & -\varepsilon \\ \varepsilon & 1 \end{pmatrix}\begin{pmatrix} e^{i\theta} & 0 \\ 0 & e^{-i\theta} \end{pmatrix}\begin{pmatrix} 1 & \varepsilon \\ -\varepsilon & 1 \end{pmatrix}\Bigg|_{\varepsilon=0}$$

$$= 2 \sin\theta \frac{\partial}{\partial y_4},$$

$$c_*(J_2) = -2 \sin\theta \frac{\partial}{\partial y_3}.$$

Therefore

$$c_*(dg) = |e^{i\theta} - e^{-i\theta}|^2 \frac{1}{2\pi}\, \omega \Lambda\, d\theta,$$

and we obtain the formula. □

3. Plancherel Formula for G

Theorem. *Let dg be the Haar measure on G giving total mass 1 to G, then:*

$$\varphi(e) = \sum_{n=0} n \operatorname{tr} T_n(\varphi) \quad \textit{for every } \varphi,\ C^\infty \textit{ function on } G.$$

Proof. We have

$$\operatorname{tr} T_n(\varphi) = \int_G \operatorname{tr} T_n(g)\varphi(g)\, dg$$

$$= \frac{1}{4\pi}\int_0^{2\pi} |e^{i\theta} - e^{-i\theta}|^2 \operatorname{tr} T_n\left(g\begin{pmatrix} e^{i\theta} & 0 \\ 0 & e^{-i\theta} \end{pmatrix}g^{-1}\right)\varphi\left(g\begin{pmatrix} e^{i\theta} & 0 \\ 0 & e^{-i\theta} \end{pmatrix}g^{-1}\right) dg\, d\theta$$

$$= \frac{1}{4\pi}\int_0^{2\pi} (e^{-i\theta} - e^{i\theta})(e^{in\theta} - e^{-in\theta}) \int_G \varphi\left(g\begin{pmatrix} e^{i\theta} & 0 \\ 0 & e^{-i\theta} \end{pmatrix}g^{-1}\right) dg\, d\theta.$$

Consider

$$(k\varphi)(\theta) = (e^{-i\theta} - e^{i\theta}) \int_G \varphi\left(g\begin{pmatrix} e^{i\theta} & 0 \\ 0 & e^{-i\theta} \end{pmatrix} g^{-1}\right) dg\, d\theta.$$

Then

$$(k\varphi)'(0) = -2i\varphi(e).$$

Let us calculate:

$$\begin{aligned}
\sum_{n>0} n \operatorname{tr} T_n(\varphi) &= \frac{1}{4\pi} \sum_{n>0} \int_0^{2\pi} (k\varphi)(\theta) n(e^{in\theta} - e^{-in\theta})\, d\theta \\
&= \frac{1}{2\pi} \sum_{n\in\mathbf{Z}} \int_0^{2\pi} (k\varphi)(\theta) n(e^{in\theta} - e^{-in\theta})\, d\theta \\
&= \frac{1}{4\pi} \sum_{n\in\mathbf{Z}} \int_0^{2\pi} (k\varphi)(\theta) n e^{in\theta}\, d\theta \\
&\quad \text{as } (k\varphi)(\theta) \text{ is odd,} \\
&= \frac{1}{4\pi} \sum_{n\in\mathbf{Z}} i \int_0^{2\pi} (k\varphi)'(\theta) e^{in\theta}\, d\theta \\
&= \frac{i}{2} (k\varphi)'(0) = \varphi(e). \quad \square
\end{aligned}$$

q.e.d.

4. Kirillov Character Formula

We want to prove the formula (3.15)

$$\frac{e^{irt} - e^{-irt}}{2it} = \int_{x_1^2+x_2^2+x_3^2=r^2} e^{ix_3 t} \frac{\sigma_r}{2\pi}.$$

Let μ be the surface measure of S_r. At the point $(0, 0, r) = rJ_3$,

$$\left|\mu\left(\frac{\partial}{\partial x_1} \wedge \frac{\partial}{\partial x_2}\right)\right| = 1.$$

The tangent vector generated by the infinitesimal action of J_1 at the point rJ_3 is given by $[J_1, rJ_3] = 2rJ_2$, while the tangent vector generated by the infinitesimal action of J_2 is $-2rJ_1$. By definition,

$$\sigma\left(2r\frac{\partial}{\partial x_2}, -2r\frac{\partial}{\partial x_1}\right) = (rJ_3^*, [J_1, J_2]).$$

Thus $\sigma/2\pi = (1/4\pi r)\mu$. By considering spherical coordinates on S_r, the second member is:

$$\frac{1}{4\pi r} \iint e^{i(r\cos\varphi t)} r^2 \sin\varphi\, d\varphi\, d\theta,$$

which leads immediately to the above formula.

5. The Poisson–Plancherel Formula

Define, for φ a Schwartz function on $\mathfrak{g}$,

$$(K\varphi)(t) = \int_{S_t} \varphi \frac{\sigma}{2\pi} \qquad \text{for } t > 0,$$

$$= -\int_{S_{|t|}} \varphi \frac{\sigma}{2\pi} \quad \text{for } t < 0.$$

In spherical coordinates,

$$(K\varphi)(t) = \frac{t}{4\pi} \iint \varphi(t \cos\theta \sin\varphi, t \sin\theta \sin\varphi, t \cos\varphi) \sin\varphi \, d\varphi \, d\theta.$$

It is immediate to see that $(K\varphi)(t)$ can be extended as a C^∞-odd function of t and that:

$$\int \varphi(x_1, x_2, x_3) \, dx_1 \, dx_2 \, dx_3 = 2\pi \int_{-\infty}^{\infty} t(K\varphi)(t) \, dt.$$

Define

$$(M_t, \varphi) = \left(\frac{\partial}{\partial t} K\varphi\right)(t).$$

It follows from the expression of $K\varphi(t)$ in spherical coordinates that $(M_0, \varphi) = \varphi(0)$.

Recall that the set $\mathfrak{g}_G$ consists of all the spheres S_t with

$$t \in 2\pi\mathbf{Z} \qquad (t \geq 0).$$

Define

$$(n_G, \varphi) = \sum_{a \in 2\pi\mathbf{Z}} (M_a, \varphi).$$

Recall (3.22) that Θ_a is the G-invariant function on $\mathfrak{g}^*$ such that

$$\Theta_a\begin{pmatrix} i\lambda & 0 \\ 0 & -i\lambda \end{pmatrix} = \tfrac{1}{2}(e^{i\lambda a} + e^{-i\lambda a})$$

and that v_G is the G-invariant generalized function $\sum_{a \in 2\pi\mathbf{Z}} \Theta_a$. Let us verify that the couple (n_G, v_G) verifies the Poisson–Plancherel formula. This will follow from the

Proposition. (a) *n_G is a distribution of support $\mathfrak{g}_G$ and n_G coincides with ∂ in a neighborhood of* 0.

(b) $\hat{M}_a = \Theta_a$,

(c) *If dg is the Haar measure on G giving total mass* 1 *to G, dX the Euclidean measure on $\mathfrak{g}$ such that $dg = j(X)\,dX$, df the dual measure on $\mathfrak{g}^*$, then, for φ a Schwartz function on $\mathfrak{g}^*$,*

$$\int_{\mathfrak{g}^*} \varphi(f) v_G(f) \, df = \sum_{n \in N} n \int_{S_n} \varphi \frac{\sigma_n}{2\pi}.$$

Proof. (a) is clear.

(b) We have to verify that $(M_a, e^{i\langle \xi, X\rangle}) = \Theta_a(\xi)$. By G-invariance, it is sufficient to verify this for $\xi = \begin{pmatrix} i\lambda & 0 \\ 0 & -i\lambda \end{pmatrix}$, i.e. that $(M_a, e^{i\lambda x_3}) = \frac{1}{2}(e^{i\lambda a} - e^{-i\lambda a})$. But

$$K(e^{i\lambda x_3})(a) = \int_{S_a} e^{i\lambda x_3} \frac{\sigma_a}{2\pi}$$
$$= \frac{e^{i\lambda a} - e^{-i\lambda a}}{2i\lambda}, \quad \text{by 4.}$$

thus

$$M_a(e^{i\lambda x_3}) = \frac{\partial}{\partial a} K(e^{i\lambda x_3})(a) = \tfrac{1}{2}(e^{i\lambda a} + e^{-i\lambda a}).$$

(c) We have seen that $dX = dx_1\, dx_2\, dx_3/2\pi^2$, thus $df = d\xi_1\, d\xi_2\, d\xi_3/4\pi$.
By definition of v_G:

$$\int_{\mathfrak{g}^*} \varphi(f) v_G(f)\, df = \sum_{a \in 2\pi\mathbf{Z}} \int \Theta_a(f)\varphi(f)\, df$$
$$= \sum_{a \in 2\pi\mathbf{Z}} \frac{1}{2} \int_{-\infty}^{\infty} tK(\Theta_a \varphi)(t)\, dt$$
$$= \sum_{a \in 2\pi\mathbf{Z}} \frac{1}{2} \int_{-\infty}^{\infty} \frac{e^{iat} + e^{-iat}}{2} t(K\varphi)(t)\, dt$$
$$= \sum_{n \in \mathbf{Z}} \frac{1}{2} \int_{-\infty}^{\infty} \frac{e^{2i\pi nt} + e^{-2i\pi nt}}{2} t(K\varphi)(t)\, dt$$
$$= \sum_{n \in \mathbf{Z}} \frac{1}{2} \int_{-\infty}^{\infty} e^{2i\pi nt} t(K\varphi)(t)\, dt$$

as $t(K\varphi)(t)$ is an even function of t

$$= \sum_{n \in \mathbf{Z}} \tfrac{1}{2} n(K\varphi)(n)$$

from the usual Poisson summation formula,

$$= \sum_{n > 0} n \int \varphi \frac{\sigma_n}{2\pi}. \quad \square$$

q.e.d.

APPENDIX 3

Complement on the Poisson–Plancherel Formula for SL(2, R)

Recall that SL(2, $\mathbf{R}$) has two conjugacy classes of Cartan subalgebras $\mathfrak{b}$ and $\mathfrak{a}$ with

$$\mathfrak{b} = \left\{ \begin{pmatrix} 0 & t \\ -t & 0 \end{pmatrix}; t \in \mathbf{R} \right\}, \qquad \mathfrak{a} = \left\{ \begin{pmatrix} s & 0 \\ 0 & -s \end{pmatrix}; s \in \mathbf{R} \right\}.$$

We denote by $\mathcal{O}_t^d$ the conjugacy class of the element $\begin{pmatrix} 0 & t \\ -t & 0 \end{pmatrix}$, $\mathcal{O}_s^p = \mathcal{O}_{-s}^p$ the conjugacy class of the element $\begin{pmatrix} s & 0 \\ 0 & -s \end{pmatrix}$. Each orbit $\mathcal{O}$ has a canonical Liouville measure $dm_{\mathcal{O}}$. Define, for φ a C^∞ function on $\mathfrak{g}$,

$$(K_{\mathfrak{b}}\varphi)(t) = \int_{\mathcal{O}_t^d} \varphi \, dm_T \qquad \text{for } t > 0,$$

$$= -\int_{\mathcal{O}_t^d} \varphi \, dm_t \qquad \text{for } t < 0,$$

$$(K_{\mathfrak{a}}\varphi)(s) = \int_{\mathcal{O}_s^p} \varphi \, dm_s.$$

We have the integration formula:

$$\int \varphi(x_1, x_2, x_3) \, dx_1 \, dx_2 \, dx_3 = 4\pi \int_{-\infty}^{\infty} t(K_{\mathfrak{b}}\varphi)(t) \, dt + 4\pi \int_0^{\infty} |s|(K_{\mathfrak{a}}\varphi)(s) \, ds.$$

Define:

$$(M_t, \varphi) = \frac{\partial}{\partial t}(K_{\mathfrak{b}}\varphi)(t) \quad \text{for } t \neq 0.$$

It is not difficult to see that (M_t, φ) can be extended to a continuous function of t and that $(M_0, \varphi) = \varphi(0)$.

Let us now define n_G and v_G. Recall that

$$\mathfrak{g}_G = \bigcup_{a \in 2\pi\mathbf{Z}} \mathcal{O}_a^d,$$

$$\mathfrak{g}_G^* = \left(\bigcup_{n \in \mathbf{Z} - \{0\}} \mathcal{O}_n^d \right) \cup \bigcup_{s \neq 0} \mathcal{O}_s^p.$$

Define

$$(n_G, \varphi) = \sum_{a \in 2\pi\mathbf{Z}} (M_a, \varphi).$$

Recall that Θ_a is the G-invariant function on $\mathfrak{g}^*$ such that

$$\Theta_a \begin{pmatrix} 0 & \lambda \\ -\lambda & 0 \end{pmatrix} = e^{ia\lambda}$$

$$\Theta_a \begin{pmatrix} s & 0 \\ 0 & -s \end{pmatrix} = e^{-|as|},$$

and that

$$v_G = \sum_{a \in 2\pi\mathbf{Z}} \Theta_a.$$

Define

$$(n_G, \varphi) = \sum_{a \in 2\pi\mathbf{Z}} (M_a, \varphi).$$

Lemma.

$$(v_G, \varphi\, df) = 4\pi \sum_{n \in \mathbf{Z}} |n| \left(\int_{\mathcal{O}_n^d} \varphi \frac{\sigma}{2\pi} \right) + 4\pi \int_0^\infty s \coth \pi s \left(\int_{\mathcal{O}_s^p} \varphi \frac{\sigma}{2\pi} \right) ds.$$

Proof. By definition

$$\begin{aligned}\int v_G(f)\varphi(f)\, df &= \sum_{a \in 2\pi\mathbf{Z}} \int \Theta_a(f)\varphi(f)\, df \\ &= \sum_{a \in 2\pi\mathbf{Z}} 4\pi \int_{-\infty}^{\infty} tK_{\mathfrak{b}}(\Theta_a \varphi)(t)\, dt + 4\pi \int_0^\infty sK_{\mathfrak{a}}(\Theta_a \varphi)(s)\, ds \\ &= \sum_{a \in 2\pi\mathbf{Z}} 4\pi \int_{-\infty}^{\infty} e^{iat} t(K_{\mathfrak{b}}\varphi)(t)\, dt + 4\pi \int_0^\infty e^{-|as|} s(K_{\mathfrak{a}}\varphi)(s)\, ds.\end{aligned}$$

Now the lemma follows from the usual Poisson summation formula on $\mathbf{R}$ ($t(K_{\mathfrak{b}}\varphi)(t)$ is a continuous function of t) and from the formula, for $s > 0$

$$\begin{aligned}\sum_{n \in \mathbf{Z}} e^{-2\pi|ns|} &= 2 \sum_{n > 0} e^{-2\pi ns} + 1 \\ &= 2 \sum_{n \geq 0} e^{-2\pi ns} - 1 \\ &= \frac{2}{1 - e^{-2\pi s}} - 1 \\ &= \frac{1 + e^{-2\pi s}}{1 - e^{-2\pi s}} \\ &= \coth \pi s.\end{aligned}$$

To prove the conjecture, it would remain to prove that

$$\hat{M}_a = \Theta_a.$$

This follows from Rossmann's formula [51] and the recurrence relation [25] for discrete series constants. □

BIBLIOGRAPHY

[1] M. F. Atiyah and R. Bott. A Lefschetz fixed point formula for elliptic complexes, II. *Ann. Math.*, **88**, 451–491 (1968).

[2] M. F. Atiyah and I. M. Singer. The index of elliptic operators III. *Ann. Math.*, **87**, 546–604 (1968).

[3] L. Auslander and B. Kostant. Polarizations and unitary representations of solvable groups. *Invent. Math.*, **14**, 255–354 (1971).

[4] M. W. Baldoni-Silva. The unitary dual of Sp(n, 1) $n \geq 2$. *Duke Math. J.*, **48**, 549–583 (1981).

[5] M. W. Baldoni-Silva and D. Barbasch. The unitary spectrum for real rank one groups. Preprint, 1981.

[6] V. Bargmann. Irreducible unitary representations of the Lorentz group. *Ann. Math.*, **48**, 568–640 (1947).

[7] N. Berline and M. Vergne. Fourier transform of orbits of the co-adjoint representations. To appear in *Proceedings of the Utah Conference on "Representations of Reductive Groups,"* Park City, April 1982, Birkhaüser: Boston.

[8] N. Berline and M. Vergne. Classes caracteristiques equivariantes. Formule de localisation en cohomologie equivariante. *Compte-Rendus a l'Académie des Sciences, Paris.* t. 295, 15 Novembre 1982, pp. 539–541.

[9] P. Bernat *et al. Représentations des groupes de Lie résolubles.* Dunod: Paris, 1972.

[10] R. Bott. Vector fields and characteristic numbers. *Mich. Math. J.*, **14**, 231–244 (1967).

[11] J. Y. Charbonnel. La formule de Plancherel pour un groupe résoluble connexe, II. *Math. Ann.*, **250**, 1–34 (1980).

[12] A. Connes and H. Moscovici. The L^2-index theorem for homogeneous spaces of Lie groups. *Ann. Math.*, **115**, 291–330 (1982).

[13] J. Dixmier. Represéntations irreductibles des algèbres de Lie nilpotentes. *Anais Da Academia Brasileira de Ciencias*, **35**, 491–519 (1963).

[14] J. Dixmier. *Les C*-algèbres et leurs Représentations*, Gauthier-Villars: Paris, 1964.

[15] J. Dixmier. *Algèbres Enveloppantes*, Gauthier-Villars: Paris, 1974.

[16] M. Duflo. Représentations unitaires irreductibles des groupes simples complexes de rang deux. *Bull. Soc. Math. France*, **107**, 55–96 (1979).

[17] M. Duflo. *Construction de Representations Unitaires d'un Groupe de Lie.* C.I.M.E.: Cortona, 1980, Liguori editore, Napoli, 1982.

[18] M. Duflo. Théorie de Mackey pour les groupes algébriques. Preprint, 1981.

[19] M. Duflo. On a conjecture of M. Vergne on the Poisson–Plancherel formula: the case of complex groups. Preprint, 1981.

[20] M. Duflo. Representations unitaires des groupes de Lie et methode des orbites. *6ᵉ-Congrès du Groupement des Mathematiciens d'Expression Latine-Luxembourg, 7–12 Septembre 1981.*

[21] M. Duflo and M. Vergne. Une propriété de la représentation coadjointe d'une algèbre de Lie. *C. R. Acad. Sci., Paris*, **268**, 583–585 (1969).

[22] J. Duistermaat and G. Heckman. On the variation in the cohomology form of the reduced phase space. To appear in *Invent. Math.*

[23] V. Guillemin and S. Sternberg. *Geometric Asymptotics. Mathematical Surveys*, 14. Providence, 1977.

[24] Harish-Chandra. Plancherel formula for the 2 × 2 real unimodular group. *Proc. Nat. Acad. Sci.*, **38**, 337–341 (1952).

[25] Harish-Chandra. Discrete series for semi-simple Lie groups, I–II. *Acta Math.*, **113**, 241–318 (1965); **116**, 1–111 (1966).

[26] Harish-Chandra. Harmonic analysis on real reductive groups, III. *Ann. Math.*, **104**, 117–201 (1976).

[27] R. Herb. Characters of averaged discrete series on semi-simple real Lie groups. *Pacific J. Math.*, **80**, 169–177 (1979).

[28] R. Herb. Fourier inversion and the Plancherel theorem for real semi-simple Lie groups. *Amer. J. Math.*, **104**, 9–58 (1982).

[29] R. Herb. Discrete series characters and Fourier inversion on semi-simple real Lie groups. Preprint, 1980.

[30] R. Herb. Discrete series characters and Fourier inversion. Preprint, 1980.

[31] R. Herb. Fourier inversion and the Plancherel theorem. In *Non-Commutative Harmonic Analysis and Lie Groups*, Luminy, 1980. Springer Lecture Notes in Maths, 880. Springer-Verlag: New York.

[32] R. Herb. The Plancherel theorem for semi-simple groups without compact Cartan subgroups. To appear in *Non-Commutative Harmonic Analysis and Lie Groups*, Luminy, 1982. Springer Lecture Notes in Maths. Springer-Verlag: New York.

[33] M. S. Khalgui. Sur les caractères des groupes de Lie a radical co-compact. *Bull. Soc. Math. France*, **109**, 331–372 (1981).

[34] M. S. Khalgui. Sur les caractères des groupes de Lie. *J. Funct. Anal.*, **47**, 64–77 (1982).

[35] A. A. Kirillov. Unitary representations of nilpotent Lie groups. *Usp. Mat. Nauk*, **17**, 57–110 (1962).

[36] A. A. Kirillov. Characters of unitary representations of Lie groups. *Funct. Anal. App.*; **2-2**, 40–55 (1967).

[37] A. A. Kirillov. *Elements of the Theory of Representations*, M.I.R.: Moscow, 1974; Springer-Verlag: Berlin, New York, 1976.

[38] A. Knapp and B. Speh. Irreducible unitary representations of SU(2, 2). *J. Funct. Anal.*, **45**, 41–73 (1982).

[39] Y. Kosmann and S. Sternberg. Conjugaison des sous-algèbres d'isotropie, *C. R. Acad. Sci., Paris*, **279**, 777–779 (1974).

[40] B. Kostant. Quantization and unitary representation theory. I. Prequantization. *Lecture in Modern Analysis and Applications, III*. Lecture Notes in Math., Vol. 170, Springer-Verlag: Berlin, 1970, pp. 87–208.

[41] S. Lang. SL(2, **R**). Addison-Wesley: Reading, Mass.: 1975.

[42] R. Lipsman, Characters of Lie Groups, II: Real polarizations and the orbital integral character formula. *J. d'Anal. Math.*, **74**, 329–387 (1961).

[43] G. W. Mackey. Unitary representations of group extensions. *Acta Math.*, **99**, 265–311 (1958).

[44] C. Moeglin. Ideaux bilateres des algebres enveloppantes. *Bull. Soc. Math. France*, **108**, 143–186 (1980).

[45] C. Moeglin. Ideaux primitifs des algebres enveloppantes. *J. Math. Pures et Appl.*, **59**, 265–336 (1980).

[46] C. Moeglin and R. Rentschler. Orbites d'un groupe algèbrique dans l'espace des ideaux rationnels d'une algèbre enveloppante. *Bull. Soc. Math. France*, **109**, 403–426 (1981).

[47] J. Von Neumann. Die Eindeugtigkeit des Schrödingerschen operatoren, *Math. Ann.*, **104**, 570–578 (1931).

[48] R. Penney. Lie cohomology of representations of nilpotent Lie groups and holomorphically induced representations. *Trans. Amer. Math. Soc.*, **261**, 33–51 (1980).

[49] L. Pukanszky. Unitary representations of solvable Lie groups. *Ann. l'E.N.S., Paris*, **4**, 457–608 (1971).

[50] J. Rosenberg. Realizations of square integrable representations of unimodular Lie groups in L^2-cohomology spaces. *Trans. Amer. Math. Soc.*, **261**, 1–32 (1980).

[51] W. Rossmann. Kirillov's character formula for reductive Lie groups. *Invent. Math.*, **48**. 207–220 (1978).

[52] W. Schmid. On a conjecture of Langlands, *Ann. Math.*, **93**, 1–42 (1971).

[53] W. Schmid. On the characters of the discrete series. *Invent. Math.*, **30**, 47–144 (1975).

[54] D. Shelstad. Orbital integrals and a family of groups attached to a real reductive group. *Ann. l'E.N.S., Paris*, **12**, 1–31 (1979).

[55] D. Shelstad. Embeddings of L-groups. Preprint.

[56] D. Shelstad. *L*-indistinguishability for real groups. Preprint.

[57] D. Shelstad. Base change and matching theorem for real groups. In *Non-Commutative Harmonic Analysis and Lie Groups*, Luminy, 1980. Springer Lecture Notes in Maths., 880. Springer-Verlag: New York.

[58] J. M. Souriau. Structure des systèmes dynamiques. In *Maitrise de Mathématiques*. Dunod: Paris, 1970.

[59] B. Speh. Some results on principal series representations of GL(*n*, **R**). M.I.T. thesis, 1977.

[60] B. Speh. The unitary dual of GL(3, **R**) and GL(4, **R**), *Math. Ann.*, **259**, 113–133 (1981).

[61] M. Vergne. A Plancherel formula without group representations. *Proceedings of the O.A.G.R. Conference*: Bucharest, 1980, to appear in *Operator Algebras and Group Representations*, Pitman, London.

[62] M. Vergne. A Poisson–Plancherel formula for semi-simple Lie groups. *Ann. Math.*, **115**, 639–666 (1982).

[63] D. Vogan. Singular unitary representations. In *Non-Commutative Harmonic Analysis and Lie Groups*, Luminy, 1980. Springer Lecture Notes in Maths., 880. Springer-Verlag: New York.

[64] H. Weyl. Theorie der darstellung kontinuerlichen halbenfachen groupen durch lineare transformationen, I, II, III. *Math. Z.*, **23**, 271–309 (1925); **24**, 328–376 (1926); **24**, 377–395 (1926).

Conservation Laws and Their Application in Global Differential Geometry

KAREN UHLENBECK*

Most of E. Noether's mathematical research was in the field of algebra, or related to it. Outside this work in algebra is one very famous theorem of Noether's which is stated in every book on classical mechanics, as well as in many texts on more recently developed physical theories [20]. This well-known theorem applies to the following situation: an (action) integral in the calculus of variations has continuous one-parameter groups of symmetries (or the infinitesimal version of these). Then Noether's theorem states that to each such symmetry is associated a conservation law, or first integral† of the Euler–Lagrange equations for the integral.

This theorem is important in classical mechanics and general relativity and has played a basic conceptual and philosophical role in the more recent developments in particle physics and field theory. The supersymmetric models in unified field theories are attempts to extend the ideas of Noether's theorem to situations in which there are technical difficulties in defining action integrals.

This theorem of Noether's is certainly more important in physical theories than in mathematics. For one reason, the role of the calculus of variations is more central in physics. The importance of the calculus of variations in geometry has grown recently. When mathematicians have integrals or conservation laws in a variational problem, they often find and use them without emphasizing the connection with symmetries through Noether's theorem. Despite this, these conservation laws can be important in differential geometry. Moreover, mathematicians do associate first integrals with continuous group actions, and one of the most beautiful and active subjects developed in recent years in mathematics

* Department of Mathematics, University of Illinois, Box 4348, Chicago, IL 60680, U.S.A.

† First integral and conserved quantity are used interchangeably in this lecture. The use of the phrase "conservation law" is less confusing than "integral" in the calculus of variations, particularly in multiple integral problems. Integral can also mean the functional, a functional constraint, or worst of all, an integral manifold for an exterior system of partial differential equations!

is the study of completely integrable systems in which the integrals are not obviously due to the existence of symmetries [17], [19]. This is not the subject of this lecture.

I cannot adequately explain the importance of Noether's theorem in modern physics. The goal of this lecture is only to make Noether's theorem more familiar to mathematicians. After an elementary outline of facts about the calculus of variations, Noether's theorem is stated for Lagrangian mechanical systems, and the standard examples plus a few more are given. Then the theorem in its proper, relativistic form is given, along with examples of the geometric variational problems of current interest. Many of these problems are related to the variational formulation of problems in general relativity, which were the original motivation for and application of Noether's theorem. Also, many of these geometric problems possess intrinsic symmetries, and therefore conservation laws.

First let me recall for the non-experts the basic ideas in the calculus of variations. The elementary theorem of advanced calculus is familiar.

Assume $f\colon X^n \to \mathbf{R}$ is a differentiable function on an open region in $\mathbf{R}^n$, or an n-dimensional manifold. If f takes on a local minimum or maximum at $x_0 \in X^n$, then

$$df(x_0) = \left(\frac{\partial f}{\partial x^1}(x_0), \ldots, \frac{\partial f}{\partial x^n}(x_0)\right) = \mathbf{0}.$$

Associated with this theorem is the definition:

If $df(x_0) = 0$, then x_0 is a *critical point* of f.

Further properties of x_0 can be discovered by looking at the matrix of second derivatives. However, Noether's theorem applies to all critical points and is not related to properties of maxima and minima. Recall also, that when constraints are imposed, Lagrange multipliers appear in the equations for a critical point.

In the calculus of variations, the finite dimensional domain space X^n is replaced by an infinite dimensional domain space of functions $X^\infty = C^\infty(M, \mathbf{R}^k)$, or a global space of maps $C^\infty(M, N)$ between two finite dimensional manifolds M and N. The functions f must be given in the form of integrals (action integrals in physics). If $X^\infty = \{s \in C^\infty(M, N)\}$, then the action integral or functional is

$$f(s) = \int_M L(x, s(x), ds(x))\, d\mu.$$

Here $L\, d\mu$ is called the *Lagrangian* and is given invariantly as a map from $T^*M \otimes TN$ to measures (n-forms) on M. However, for the purposes of this lecture we need give only local descriptions, where coordinates are chosen on M and N, and

$$L(x, s(x), ds(x))\, d\mu = L\left(x^\alpha, s^i(x), \frac{\partial s^i}{\partial x^\alpha}(x)\right) dx^1 \cdots dx^n.$$

In coordinates, the Lagrangian L is a real-valued function of variables $(x, s, ds) = \{x^\alpha, q^i, p^i_\alpha\}$. where Greek indices α are domain indices, $1 \le \alpha \le n = \dim M$, and

Latin indices i are range indices, $1 \leq i \leq m = \dim N = \dim(\text{range})$. Critical points are defined as for $\mathscr{E}$ a function on finite dimensional domain. Classically one wrote $\delta\mathscr{E}$ instead of $d\mathscr{E}$, and variations, which should be elements of the tangent space to $C^\infty(M, N)$, were written δs. Critical points were classically called *stationary* curves or functions. The modern approach is described in full detail in Palais' book *Foundations of Global Non-linear Analysis* [21].

The main lemma in the calculus of variations identifies critical points of the action integral $\mathscr{E} = \int L\, d\mu$ with solutions of the Euler–Langrange equations for $\mathscr{E}$, a second-order system of differential equations when $\dim M = 1$, or a system of partial differential equations in the variables of the domain M in the general case. Recall $x = \{x^\alpha\} \in M$, $s(x) = \{q^i\} \in N$ and $ds = \{p^i_\alpha\} \in T^*_x M \times T_q N$ in local coordinates. Then the Euler–Lagrange equations for $\int L\, d\mu = \int L\, dx$ are the n equations, $1 \leq i \leq n$,

$$\sum_{\alpha=1}^{n} \frac{\partial}{\partial x^\alpha}\left[\frac{\partial L}{\partial p^i_\alpha}(\ , s, ds)\right] - \frac{\partial L}{\partial q^i}(\ , s, ds) = 0.$$

If constraints on s are present, then Lagrange multipliers appear.

To illustrate, let $M \subset \mathbf{R}^n$ be a domain, and $N = \mathbf{R}^m$ the range. The simplest functional is the energy or Dirichlet integral, $L(x, q, p) = \frac{1}{2}\sum_{\alpha=1}^n \sum_{i=1}^m (p^i_\alpha)^2$.

$$\mathscr{E}(s) = \frac{1}{2}\int_M |ds|^2\, d\mu = \frac{1}{2}\int_M \sum_{i=1}^{m}\sum_{\alpha=1}^{n}\left(\frac{\partial s^i}{\partial x^\alpha}\right)^2 dx^1 \cdots dx^n.$$

Then if $\psi = \delta s$,

$$\delta\mathscr{E} = d\mathscr{E}(\psi) = \int_M (ds, d\psi)\, d\mu = \int_M \sum_{i=1}^{m}\sum_{\alpha=1}^{n} \frac{\partial s^i}{\partial x^\alpha}\frac{\partial \psi^i}{\partial x^\alpha} dx^1 \cdots dx^n$$

$$= -\int_M \sum_i \Delta s^i \psi^i\, dx^1 \cdots dx^n.$$

The Euler–Lagrange equations are

$$\sum_{\alpha=1}^{n}\left(\frac{\partial}{\partial x^\alpha}\right)^2 s^i = \Delta s^i = 0, \qquad i = 1, 2, \ldots, m.$$

If M is a Riemannian manifold, the metric on M is put in to compute $|ds|^2\, d\mu$ which gives us a generalized energy integral on any Riemannian manifold, as well as a Laplacian Δ.

If the constraint $\int_M |s|^2\, d\mu = 1$ is added, the equations become, with the addition of the Langrange multiplier λ

$$\Delta s^i + \lambda s^i = 0.$$

Here $\lambda = \int_M |ds|^2\, d\mu \in \mathbf{R}$ if $s|\partial M = 0$ and $\int_M |s(x)|^2\, d\mu = 1$.

If the constraint $|s(x)|^2 = 1$ for all x is added, the equations acquire another type of Lagrange multiplier

$$\Delta s^i + \lambda(x) s^i = 0.$$

Here the function $\lambda(x) = |ds|^2(x)$ can be computed from the constraint.

Both constraints and boundary conditions are essential ingredients in the general theory of the calculus of variations. However, Noether's theorem is insensitive to boundary conditions, when properly stated. Also, for "symmetric" constraints, the conservation laws are exactly as when no constraints are present. This simplifies our discussion.

1. Noether's Theorem in Langrangian Mechanics

In Lagrangian mechanics, motion of a point particle is given by a path $s(t)$ in a manifold N. The particle "obeys" the principle of least action, or $s(t)$ is a solution curve for the Euler–Lagrange equations of the action integral

$$\mathcal{A}(s) = \int_a^b L(t, s(t), s'(t))\, dt.$$

The Lagrangian L is kinetic energy minus potential energy, but in an extended sense can be any function $L(t, q, p)$, where $(t, q) \in \mathbf{R} \times N$, $p \in T_q N$, the tangent space to N at q. In local coordinates, $(t, q, p) \in \mathbf{R} \times \mathbf{R}^m \times \mathbf{R}^m$.

A continuous symmetry for a mechanics problem is given by a one-parameter family of diffeomorphisms which are differentiable in the parameter ρ at $\rho = 0$. (In fact, in this lecture we *always* assume everything is smooth). So $u_\rho\colon \mathbf{R} \times N \to \mathbf{R} \times N$, $\rho \in (-\varepsilon, \varepsilon)$ and $u_0 =$ identity. If $s\colon [a, b] \to N$, then for small ρ, we can define $u_\rho(s)$ as the function whose graph is $u_\rho(t, s(t))$. Note that the domain may not be invariant under the diffeomorphisms.

Definition. The integral $\mathcal{A}$ *is invariant* under the one-parameter family of symmetries if $\mathcal{A}(s) = \mathcal{A}(u_\rho(s))$ for all domains $[a, b] \in \mathbf{R}$ and all $s \in C^\infty([a, b], N)$, $\rho \in (-\varepsilon, \varepsilon)$. We say $\mathcal{A}$ is *infinitesimally invariant* if $d/d\rho|_{\rho=0}\, \mathcal{A}(u_\rho(s)) = 0$.

Clearly $\mathcal{A}$ is infinitesimally invariant if it is invariant. Just as obviously, it is a good deal easier to check invariance than infinitesimal invariance.

An integral constraint is preserved according to the same definition. A continuous constraint $(t, s(t)) \subseteq \tilde{N} \subset \mathbf{R} \times N$ is preserved if $u_\rho(\tilde{N}) \in \tilde{N}$.

Definition. A *first integral* (*or equivalently, conserved quantity*) for a system of second-order ordinary differential equations on N is a function

$$J\colon \mathbf{R} \times TN \to \mathbf{R} \qquad (J = J(t, q, p))$$

such that for any solution $s(t)$ to the equations $(d/dt)J(\ , s, s') = 0$. Equivalently, J is constant on solution curves, or $J(t, s(t), s'(t)) \equiv J(0, s(0), s'(0)) \equiv$ constant.

Noether's Theorem [1], [2]. *If an integral $\mathcal{A}$ is infinitesimally invariant under a family of diffeomorphisms u_ρ, then there is a unique first integral associated with this invariance. If constraints are present and preserved by the diffeomorphism u_ρ, the same first integrals occur.*

In terms of local coordinates $(t, s, s') = (t, q, p)$ there is an infinitesimal vector field on $\mathbf{R} \times N$ associated with u_ρ which has projections τ_ρ along $\mathbf{R}$ and V_ρ along N.

$$\left.\frac{\partial}{\partial \rho}\right|_{\rho=0} u_\rho(t, q) = (\tau_\rho(t, q, V_\rho(t, q)) \in \mathbf{R} \times T_q N.$$

In terms of this vector field

$$J(t, q, p) = L(t, q, p)\tau_\rho(t, q) + \sum_i \frac{\partial L}{\partial q^i}(t, q, p)[V^i_\rho(t, q) - p^i \tau_\rho(t, q)].$$

This is much easier to follow after a quick run-through of the standard examples.

Whenever the Langrangian $L = L(q, p)$ does not depend on the time parameter t, the action integral $\mathscr{E}$ is invariant under the one parameter *group* (note u_ρ does not have to be a group) $u_\rho(t, q) = (t + \rho, q)$, the group of time translations. Here $\tau_\rho(t, q) = 1$ and $v_\rho = 0$. The conserved quantity is *energy*

$$E(q, p) = \sum_i \frac{\partial L}{\partial p^i}(q, p)p^i - L(q, p).$$

When $N = \mathbf{R}^m$ and q does not appear in the formula for L, then $\mathscr{E}$ is invariant under the m translations $u_{k,\rho}(t, q) = (t, q + \rho e_k)$ (where $e^j_k = \delta^j_k$, $k = 1, 2, \ldots, m$). The conserved quantity is the *linear momentum vector*

$$m = \{m_k(t, q, p)\} = \left\{\frac{\partial L}{\partial p^k}(t, q, p)\right\}.$$

Suppose, again, that $N = \mathbf{R}^m$ and $L = L(|q|, |p|)$. Then $\mathscr{E}$ is invariant under orthogonal rotations (infinitesimally represented by $m \times m$ skew matrices) in $\mathbf{R}^m$ (sometimes this may be partially true). Then *angular momentum* is conserved.

$$\mu_{kj}(t, p, q) = m_k(t, q, p)q^j - m_j(t, q, p)q^k = (m \wedge q)_{kj}.$$

A simple example illustrates this. Let $N = \mathbf{R}^3$ and

$$\mathscr{E}(s) = \int_a^b \tfrac{1}{2}m|s'(t)|^2 - mKs^3(t)\,dt$$

or $L(t, q, p) = \frac{1}{2}m\sum_{i=1}^3 (p^i)^2 - mKq^3$. The group which leaves this integral invariant is generated by time translation, translations and rotations in the (q^1, q^2) plane. So the first integrals are energy $\frac{1}{2}\sum_{i=1}^3 m(p^i)^2 + mKq^3$, linear momentum in the plane (q^1, q^2) which is (mp^1, mp^2), and the component of angular momentum $\mu_{12} = m(p^1q^2 - q^1p^2)$. If the constraint $|s(t)|^2 \equiv 1$ (or $|q|^2 = \sum_{i=1}^3 (q^i) = 1$) is imposed, then energy and angular momentum in the (q^1, q^2) plane are still conserved, but linear momentum is not, since the constraint is preserved under rotation but not translation. Note we avoid computing the Euler–Lagrange equations. This is the action integral for the solid pendulum (motion of a particle under a gravitational force contrained to a sphere).

The classic geometric problem where first integrals are used is the case of geodesics on a Riemannian manifold N with a continuous isometry group G. Here the action integral on $s: [a, b] \to N$ is $E(s) = \frac{1}{2}\int_a^b |s'(t)|^2\, dt$, where the Riemannian metric on N is used to define $|s'(t)|^2$. This integral is invariant under time translation, which gives conservation of energy $|s'(t)|^2 \equiv E$ along orbits. A second set of integrals comes from the Lie group of isometries G action on N alone. Every element A of the Lie algebra of G maps canonically to a first integral $J_A(q, p) = (p \cdot K_A(q))$ where $K_A(q) = d/d\rho|_{\rho=0}\, e^{\rho A}(q)$ is the Killing vector field or infinitesimal action of $e^{\rho A}$ associated with A. If $N = S^{m-1}$, then $G = \mathrm{SO}(m)$ and these integrals are exactly *angular momenta*. Physicists call such integrals *generalized angular momentum integrals*. In the case that N is a symmetric space, there are enough integrals to completely analyze the geodesic flow [5]. Bott's original proof of the periodicity theorem uses this fact [17].

This simple form of Noether's theorem applies also to cases of multiple integrals where the image is a space of functions, $N = X^\infty = C^\infty(M, \mathbf{R}^m)$. The Lagrangian L itself is defined using integrals. Here we identify $s: \mathbf{R} \to C^\infty(M, \mathbf{R}^m)$ canonically with $\tilde{s}: \mathbf{R} \times M \to \mathbf{R}^m$ by $s(t)x = \tilde{s}(t, x)$. Define a Lagrangian, given a potential $V: \mathbf{R}^m \to \mathbf{R}$, by

$$L\left(s(t, s'(t)) = \frac{1}{2}\int_M (|s'(t)(x)|^2 - |d(s(t))(x)|^2 - 2V(s(t)x))\right) dx$$

$$= \frac{1}{2}\int_M \left(\left|\frac{\partial \tilde{s}}{\partial t}(t, x)\right|^2 - |d_x \tilde{s}(t, x)|^2 - 2V(\tilde{s}(t, x))\right) dx.$$

The Euler–Lagrange equations for this action integral are a system of m wave-equations (which are non-linear and coupled when $V: \mathbf{R}^m \to \mathbf{R}$ is not quadratic). Conservation of energy applies.

$$E(s(t), s'(t)) = \frac{1}{2}\int_M \left(\left|\frac{\partial \tilde{s}}{\partial t}(t, x)\right|^2 + |d_x \tilde{s}(t, x)|^2 + 2V(\tilde{s}(t, x))\right) dx$$

is constant in time for solutions of the Euler–Lagrange equations. If $M = \mathbf{R}^n$, then translational invariance $x^\alpha \to x^\alpha + \rho\delta^\alpha_\beta$ leads to conservation of quantities like $\int_{\mathbf{R}^n} \sum_{i=1}^k ((\partial \tilde{s}^i/\partial x^\beta)(t, x), (\partial \tilde{s}^i/\partial t)(t, x))\, dx$, *provided solutions decay sufficiently at infinity* to allow certain integrations by parts. Finally, if $V(s(t, x) = \tilde{V}(|s(t, x)|)$, the integral is invariant under rotations of s by elements in $\mathrm{SO}(m)$ acting on $\mathbf{R}^m$. Noether's theorem again applies (under sufficient decay assumptions at any ∞ parts of M) to give conservation of

$$\mu_{ij}(s(t), s'(t)) = \int_M \left(\frac{\partial \tilde{s}^i}{\partial t}(t, x)\tilde{s}^j(t, x) - \frac{\partial \tilde{s}^j}{\partial t}(t, x)\tilde{s}^i(t, x)\right) dx.$$

Imposing a constraint, such as $|s(x)|^2 = 1$ does not change the conservation of energy E or the "angular momentum" μ_{ij}. The above conservation of energy law and related energy estimates are extremely important in treating wave equations, whether linear or non-linear. I am not particularly aware of the importance in mathematics of conservation laws for the angular momentum-like quantities μ_{ij}, although they now appear all the time in physics.

We finish with one more application, to show the "unnatural" aspect of this formulation of Noether's theorem for partial differential equations, where a particular time variable is chosen. Here the multiple integral problem we are interested in is the Riemannian pure energy integral in a ball of radius σ

$$E_\sigma(\tilde{s}) = \frac{1}{2}\int_{|x|\le\sigma} \sum_{i=1}^{m}\sum_{\alpha=1}^{n}\left(\frac{\partial \tilde{s}^i}{\partial x^\alpha}\right)^2 dx^1\cdots dx^n$$

and the constraint $\tilde{s}(x)\in N\subset \mathbf{R}^m$ is imposed, where N is any smooth submanifold of $\mathbf{R}^m$. Critical points of this integral are *harmonic maps* $\tilde{s}$: $\{x\in\mathbf{R}^n, |x|\le\sigma\}\to N$. Let time $t=r=|x|$ and $M=S^{n-1}$. Putting in radial and spherical coordinates $(x)=(t,\theta)$, $s(t)\theta=\tilde{s}(t,\theta)$, $s(t)\in C^\infty(S^{n-1},N)$ we have

$$\ell(s) = \int_0^\sigma L(s(t), s'(t))\,dt = E_\sigma(\tilde{s}),$$

where

$$L(s(t), s'(t)) = \frac{1}{2}\int_{\theta\in S^{n-1}}\left(\left|\frac{\partial\tilde{s}}{\partial t}\right|^2(t,\theta) + \frac{1}{t^2}|d_\theta\tilde{s}(t,\theta|^2)t^{n-1}\right)d\theta.$$

Under dilation $u_\rho(t,s)=(\rho t, s(\rho t))$, we have $\ell(u_\rho s)=\rho^{n-2}\ell(s)$. *Note the constraint manifold N never enters into the argument.* An application of Noether's theorem gives a modified, energy type conservation law: the expression

$$\frac{1}{2}\int_{S^{n-1}}(|s'(t)(\theta)|^2 - t^{-2}|ds(t)(\theta)|^2)\,d\theta + \frac{1}{2}(n-2)\int_{|x|\le t}|d\tilde{s}|^2\,dx$$

is constant in time t for solutions s to the Euler–Lagrange equations. The integral on the right appears due to ρ^{n-2} in $\ell(u_\rho s)=\rho^{n-2}\ell(s)$. If $\tilde{s}$ is a solution in $|x|\le 1$, the value of the conserved quantity is zero. Replacing the identity by the weaker inequality

$$\int_{|x|\le t}|d\tilde{s}|^2\,dx \le \frac{t}{n-2}\int_{|x|=t}|d\tilde{s}|^2\,d\mu,$$

we obtain by integration in t an inequality

$$E_\sigma(\tilde{s}) \le \sigma^{n-2}E_1(\tilde{s}).$$

This *scaling inequality* is an important ingredient in recent work, joint with R. Schoen, on the regularity theory of harmonic maps [23]. A similar inequality holds in Yang–Mills theory [22]. These inequalities were *not* actually derived using Noether's theory, but were found by more wasteful *ad hoc* methods.

It is very important to remember that the converse to Noether's theorem is not true. The existence of first integrals or conserved quantities *does not* guarantee the existence of a symmetry group of transformations.

2. The General Formulation of Noether's Theorem

The application of Noether's theorem in its Lagrangian mechanical formulation to multiple integral problems is possible, but unnatural, as we saw in the last few examples. Even for the wave equations, that conservation of energy is incompatible with the Lorentz invariance of the action integrals. We now give the correct invariant formulation. Assume that $\ell(s) = \int_M L(x, s(x), ds(x))\, d\mu_M$ is an integral regarded as a function of maps $s \in C^\infty(M, N) = X^\infty$, where M and N are manifolds; μ a measure on M. A diffeomorphism $u: M \times N \to M \times N$ which is sufficiently close to the identity induces an action on $s: \Omega \to N, \Omega \subseteq M$ via the formula $u(\text{graph } s) = \text{graph}(u(s))$, just as in the case $M = [a, b]$.

Definition. A smooth family of diffeomorphisms $u_\rho: M \times N \to M \times N, \rho \in (-\varepsilon, \varepsilon)$, $u_0 = \text{identity}$, *leaves* ℓ *invariant* if for all $\Omega \subseteq M$, $s: \Omega \to N$, $\ell(s) = \ell(u_\rho(s))$. The integral ℓ is *infinitesimally invariant* if $d/d\rho|_{\rho=0}\, \ell(u_\rho(s)) = 0$.

Note that as before it may happen that domain $u_\rho(s) \neq$ domain $s(= \Omega)$.

A conservation law, or conserved quantity, is difficult to define because the term is used somewhat loosely, as well as differently in mathematics and physics. For domains of one dimension (one variable $x = t$), conservation laws associate a real valued constant (or constant vector, as in momentum vector) to each solution of the Euler–Lagrange equations. For domains M of dimension larger than one (with more than one variable $x = (x^1, \ldots, x^n)$), a conservation law assigns to every solution $s: M \to N$ of the Euler–Lagrange equations, a *divergence free* vector field J on Ω. This is called a *Noether current* in physics. Logically, it follows that "divergence free" in n dimensions extends "constant" in one dimension, if the constant is interpreted as a constant vector field on $\mathbf{R}$. To see the connection with the conserved integral quantities derived via the Lagrangian mechanics formalism, we take the case where $M = \mathbf{R} \times \tilde{M}$ actually factors. So we can write the vector field on Ω as a pair $J = (J_0, \tilde{J})$ where J_0 is the projection of J on (time) $\mathbf{R}$, and $\tilde{J}$ the projection on $\tilde{M}$ (space). Let $Q(t) = \int_M J_0(t, x)\, d\mu_M$. If $\tilde{M}$ is compact, and $\partial\tilde{M} = 0$ then

$$Q'(t) = -\int_{\tilde{M}} \operatorname{div} \tilde{J}(t, x)\, d\mu_{\tilde{M}} = 0$$

is a valid integration by parts. For global solutions s on $\Omega = [a, b] \times \tilde{M} \subseteq M$, we obtain $Q \equiv$ constant in time. Clearly we need to specify more in the general case before $\int_{\tilde{M}} \operatorname{div} \tilde{J}(t, x)\, d\mu_{\tilde{M}} = 0$ is a valid integration by parts. This should explain the origin of the qualifying statements in the first section about "behavior at ∞."

We content ourselves with a restricted definition of conservation laws. Keep in mind that this is definitely more restricted than generally used in either mathematics or physics. "Global" or "non-local" conservation laws exist and are extremely important.

Definition. A *local conservation law* for a second-order system of partial differential equations defined on maps $s: M \to N$ is a map

$$J: T^*M \otimes TN \to TM$$

such that for $s: \Omega \subseteq M \to N$ any local solutions of the system, the $J \circ ds$ is a divergence free vector field on Ω.

Noether's Theorem. *If $\mathscr{E}$ is an integral on $C^\infty(M, N)$ and $\mathscr{E}$ is infinitesimally invariant under a smooth family of diffeomorphisms $u_\rho: M \times N \to M \times N$, then there is a unique local conservation law associated with this invariance.*

To describe the map J, we turn to local coordinates $(x, q, p) \in \mathbf{R}^n \times \mathbf{R}^n \times \mathbf{R}^{n \times m}$ in $T^*M \otimes TN$. Split the vector field $(d/d\rho)|_{\rho=0}u_\rho(x, q)$ into its components on M and N:

$$\left.\frac{d}{d\rho}\right|_{\rho=0} u_\rho(x, q) = (\tau_\rho(x, q), V_\rho(x, q)) \in \mathbf{R}^m \oplus \mathbf{R}^n \simeq T_xM \oplus T_qN.$$

Recall that locally $L\,d\mu = L(x, q, p)\,d\mu$ as described in the introduction. Then the formula for the vector field J on M in coordinates is

$$J^\alpha(x, q, p) = L(x, q, p)\tau_\rho^\alpha(x, q) + \sum_{i=1}^m \frac{\partial L}{\partial p_\alpha^i}(x, q, p)\left(V_\rho^i(x, q) - \sum_\beta p_\beta^i \tau_\rho^\beta(x, q)\right).$$

Familiar conservation laws are the equivalent of conservation of energy in Lagrangian mechanics. They apply when $M = \mathbf{R}^n$ and $L = L(q, p)$ does not depend on the domain variable x of s. Then the integral is invariant under the n one-parameter families of translations $u_{\rho,\beta}(x, q) = x + \rho e_\beta$, (where $e_\beta = \delta_\beta^\alpha$). Here $\tau_\rho^\alpha = \delta_\beta^\alpha$ and $V_\rho^i = 0$.

$$J_\beta^\alpha(x, q, p) = -T_\beta^\alpha(x, q, p) = -T_\beta^\alpha(q, p) = L(q, p)\delta_\beta^\alpha - \sum_i \frac{\partial L}{\partial p_\alpha^i}(q, p)p_\beta^i.$$

This is sometimes called an *energy-momentum tensor.*

Both the non-linear wave equations and the harmonic map problem discussed in the first section are left invariant by translations for the non-linear wave equation, we obtain by writing out the divergence (here $x^0 = t$)

$$0 = -\frac{\partial}{\partial t}\left[\frac{1}{2}\left(\left|\frac{\partial s}{\partial t}\right|^2 + \left|\frac{\partial s}{\partial x}\right|^2\right) + V(s)\right] + \sum_{\alpha>0}\frac{\partial}{\partial x^\alpha}\left(\frac{\partial s}{\partial x^\alpha}\cdot\frac{\partial s}{\partial t}\right),$$

$$0 = \frac{\partial}{\partial t}\left(\frac{\partial s}{\partial x^\beta}\cdot\frac{\partial s}{\partial t}\right) - \sum_{\alpha>0}\frac{\partial}{\partial x^\alpha}\left(\frac{\partial s}{\partial x^\beta}\cdot\frac{\partial s}{\partial x^\alpha}\right) + \frac{\partial}{\partial x^\beta}\left[\frac{1}{2}\left|\frac{\partial s}{\partial x}\right|^2 + V(q) - \frac{1}{2}\left|\frac{\partial s}{\partial t}\right|^2\right].$$

These are invariant under the Lorentz group, as is philosophically required. The Lagrange–Mechanics Conservation laws of section hold by integrating these equations in $x \in \mathbf{R}^{n-1}$, assuming integration by parts in $x \in \mathbf{R}^{n-1} = \{(x^1, x^{m-1})\}$ gives no boundary term as $|x| \to \infty$.

Another application is to the *energy integral* $E(s) = \frac{1}{2}\int_\Omega |ds|^2\,dx$ defined on maps $s: \Omega \subseteq \mathbf{R}^n \to N \subseteq \mathbf{R}^m$, where the image manifold N is a constraint submanifold of $\mathbf{R}^m$. This is our last example from Section 1. Here $L = \frac{1}{2}|p|^2$, $T_\beta^\alpha = \sum_i p_\alpha^i p_\beta^i - \frac{1}{2}|p|^2\delta_{\alpha\beta}$. The conservation laws $\beta = 1, \ldots, n$ can be written out as

$$\sum_\alpha \frac{\partial}{\partial x^\alpha}\left(\frac{\partial s}{\partial x^\beta}\cdot\frac{\partial s}{\partial x^\alpha}\right) - \frac{1}{2}\frac{\partial}{\partial x^\beta}|ds|^2 = 0.$$

This is an identity on harmonic maps $s\colon \Omega \to N$ (solutions of the Euler–Lagrange equations for E). For $n = 2$, introduce a complex variable $z = x^1 + ix^2$, $\bar{z} = x^1 - ix^2$ and check that the above reduces to $\partial/\partial\bar{z}(\partial s/\partial z \cdot \partial s/\partial z) = 0$ [12]. This is a particularly well-known conservation law for harmonic maps with 2-dimensional domain. Note that when $E(s) = \frac{1}{2}\int_{R^2} |ds|^2\, dx < \infty$, we may deduce a *global conservation law* $(\partial s/\partial z \cdot \partial s/\partial z) = 0$. This leads us into the type of conservations laws mathematicians do not understand completely [7], [13].

The energy integral is a very important integral on general maps between Riemannian manifolds $s\colon M \to N$ [12] (it naturally generalizes the 1-dimensional geodesic problem $s\colon S^1 \to N$). (In physics, the integral is studied as a variety of *non-linear sigma model.*) Another type of conservation law occurs for arbitrary M, and symmetric N. To illustrate, assume $N = S^{m-1} = \{q \in \mathbf{R}^m : |q|^2 = 1\}$. The Lie group $\mathrm{SO}(m)$ acts on N and hence on $M \times N$. The Lie algebra of $\mathrm{SO}(m) = \{A \in L(R^n, R^n)\colon A = -A^T\}$. For each skew symmetric A, and every harmonic map $s\colon M \to S^{m-1} \subseteq \mathbf{R}^m$, Noether's theorem gives the identity

$$\operatorname{div}(\operatorname{grad} s \cdot As) = 0.$$

This leads to a set of equations which are locally equivalent to the Euler–Lagrange equations. Let

$$J_\alpha^{ij} = \frac{\partial s^i}{\partial x^\alpha} s^j - \frac{\partial s^j}{\partial x^\alpha} s^i = \left(\frac{\partial s}{\partial x^\alpha} \wedge s\right)^{ij}.$$

Then $J\colon M \to T^*M \otimes$ skew matrices. An alternate form of the Euler–Lagrange equations can be written

$$\operatorname{div} J = 0; \qquad (dJ)_{\alpha\beta} = [J_\alpha, J_\beta],$$

where $(dJ)_{\alpha\beta} = (\partial/\partial x^\alpha)J_\beta - (\partial/\partial x^\beta)J_\alpha$ is the exterior differential of the form J. These equations look related to the Yang–Mills formalism we will develop later, although the relationship is not completely clear [27].

We have touched on one important variational problem in differential geometry, geodesics and the extension to harmonic maps. Einstein's variational formulation of general relativity has the action integral $\mathscr{E}(g) = \int_M$ Scalar curvature $(g)\, d\mu(g)$ where the variable g is the Lorentz metric in space-time and is allowed to change, but $M = M^4$ is the fixed four manifold of space-time. The metric g is semi-definite. However, differential geometers consider the function

$$\mathscr{E}(g) = \int_M \text{Scalar curvature } (g)\, d\mu(g),$$

varying the *Riemannian* metric g on a manifold of arbitrary dimension, but fixing the volume $\int_M d\mu(g)$. If the conformal class of g is fixed, this is the *Yamabe integral*, which has an interesting history [26], [4], [15] and has lead to important problems in both elliptic non-linear equations and differential geometry. For a long time, analysts have had the idea of using such integrals to approach topological questions in the category of smooth manifolds (such as the Poincaré conjecture). The idea is that the existence of canonical metrics on every manifold restricts the topology of the possibilities: an exotic S^3 cannot, in fact, have a metric g which is a critical point for $\int_M \mathrm{Scalar}(g)\, d\mu(g)$ with $\int_M d\mu(g) = 1$ as constraint (such critical metrics are called Einstein metrics). However, this variational problem is in general badly posed, as critical points can take on neither minima or maxima. There has been greater success in finding Einstein metrics, or solutions of the Euler–Lagrange equations $\text{Ricci}(g) = \lambda g$, by other methods. S. T. Yau's solution of the Calabi conjecture [28] and Hamilton's recent work on the heat equation [14] are well-known important contributions.

Differential geometers would like for obvious reasons to understand this, and other integrals such as $\int_M |\text{Ricci}(g)|^2 \, d\mu(g)$. One connections with Noether's theorem is that all these integrals have as symmetry group the groups of diffeomorphisms. This is a notoriously difficult group to handle. (We omit describing Noether's results, because curvature is a *second-order* operator in the metric, and our simplified formalism only covers first-order Lagrangians.) Some recent ideas of Calabi on minimizing $\int_{M^{2n}} |\text{Ricci}(g)|^2 \, d\mu(g)$ for g in a fixed Kähler class look promising [8].

The Yang–Mills variational problem, which comes into differential geometry from quantum field theory, looks more complicated than these curvatures depending on metric integrals, but is actually somewhat simpler [6]. The object to be varied is a connection A in a principal bundle P over M with compact structure group G. We denote the curvature of A by $F(A)$. In physics, A is a potential, $F(A)$ a field. Fix a Riemannian metric on M and a biinvariant (trace) inner product on the Lie algebra $T_I G$. Then the Yang–Mills functional is

$$\mathscr{L}(A) = \int_M |F(A)|^2 \, d\mu(g).$$

The group which leaves this functional invariant is a large (infinite dimensional) group, the group of automorphisms of P, which is a group of sections of the associated bundle $P \times_{\text{Ad}} G$. In the local coordinates we have been using, $A = q = \{q_\alpha^{tv}\}$ where $\alpha = 1, 2, \ldots, \dim M$ are one-form indices and q_α is skew-Hermitian when $G = \text{SU}(m)$. We omit the upper, matrix indices. The local formula for curvature is

$$F(A) = \{F(A)_{\alpha,\beta}\} = \left\{\frac{\partial}{\partial x^\alpha} A_\beta - \frac{\partial}{\partial x^\beta} A_\alpha + [A_\alpha, A_\beta]\right\}.$$

In our general q, p formalism, this Lagrangian is

$$L(q, p) = \sum_{\alpha,\beta} |p_{\alpha,\beta} - p_{\beta,\alpha} + [q_\alpha, q_\beta]|^2,$$

where modifications in the norm must be made if M is an arbitrary manifold rather than $\mathbf{R}^n$.

In local coordinates, the group of symmetries can be represented as maps $u \in C^\infty(M, G)$. Since A is a connection $u(A) = uAu^{-1} + ud(u^{-1})$. To apply Noether's theorem, associate with the one-parameter family $u_p = \exp(\rho\psi)$, the vector field $v_\rho(x, q) = [\psi(x), q] - d\psi(x) = (\partial/\partial\rho)|_{\rho=0} u_\rho(q)$. Then the conservation law associated with this one-parameter subgroup is

$$\sum_{\alpha,\beta} \frac{\partial}{\partial x^\alpha} \left\langle F(A)_{\alpha,\beta}(x), [\psi(x), A_\beta(x)] - \frac{\partial\psi}{\partial x^\beta}(x) \right\rangle = 0.$$

Here $\langle \ , \ \rangle$ is the trace inner product on matrices. Also the metric on M changes the equation in the usual fashion if $M \neq \mathbf{R}^n$. It is disappointing that these are not particularly enlightening equations.

However, if $M = \mathbf{R}^n$, we obtain from Noether's theorem and the invariance under translations, the equations $\sum_\alpha (\partial/\partial x^\alpha) T^\alpha_\beta = 0$ where $\beta = 1, 2, \ldots, \dim M$ and

$$T^\alpha_\beta = 4 \sum_\gamma \langle F_{\alpha,\gamma}, F_{\gamma,\beta} \rangle - |F|^2 \delta^\alpha_\beta.$$

This can be integrated to give a scaling inequality

$$\int_{|x| \leq \sigma} |F(A)|^2 \, dx \leq \sigma^{n-4} \int_{|x| \leq 1} |F(A)|^2 \, dx,$$

which is valid on Yang–Mills fields [22]. As in the harmonic map case, this gives useful *a priori* estimates.

The full usefulness and interest of the Yang–Mills equations are only slowly becoming apparent. Taubes [25] and Sedlacek [24] have interesting existence theorems in four dimensions. Recently, Taubes' work was applied in a spectacular way by Donaldson, who shows that certain topological four manifolds cannot carry differentiable structures [11]. Atiyah and Bott have shown that the Yang–Mills variational problem is a very natural approach to studying moduli of vector bundles over Riemann surfaces [3]. There are further indications that the theory can be useful in studying the moduli space of stable vector bundles over Kähler manifolds of any dimension [16], [10].

We have shown that Noether's theorem has applications to standard variational problems in geometry: non-linear wave equations, the harmonic map problem, curvature integrals in Riemannian geometry, and the Yang–Mills functional. The fact that its role is underplayed in mathematics is as striking as its importance in physics. This is nearly as startling as the similarity of the problems of current interest in mathematics and physics.

References

[1] R. Abraham and J. Marsden. *Foundations of Mechanics.* Benjamin/Cummings: Reading, MA, 1978, pp. 285 and 479.

[2] V. I. Arnold, *Mathematical Methods of Classical Mechanics.* Graduate Texts in Mathematics, 60, Springer-Verlag; New York, 1980, pp. 88–90.

[3] M. Atiyah and R. Bott, Yang–Mills equations over Riemann surfaces. (To appear.)

[4] T. Aubin. Equations différentielles non-linéares et problème de Yamabe concernant la courbure scalaire. *J. Math. Pures Appl.*, **55** (1976), 269–296.

[5] R. Bott. The space of loops on a Lie group. *Mich. Math. J.*, **5** (1958), 35–61.

[6] J. P. Bourguignon and H. B. Lawson. Yang–Mills theory: Its physical origins and differential geometric aspects. *Seminar on Differential Geometry*, edited by S. T. Yau. Annals of Math. Studies, 102, Princeton University Press; Princeton, N.J., 1982, pp. 395–422.

[7] E. Calabi. Minimal immersions of surfaces in Euclidean spheres. *J. Diff. Geom.*, **1** (1967), 111–126.

[8] ———. Extremal Kähler metrics. *Seminar on Differential Geometry*, edited by S. T. Yau. Annals of Math Studies, 102. Princeton University Press: Princeton, N.J., 1982, pp. 395–422.

[9] S. S. Chern. On the minimal immersions of the two-sphere in a space of constant curvature. *Problems in Analysis*, edited by R. Gunning, Princeton University Press: Princeton, N.J., 1970, pp. 27–40.

[10] S. Donaldson. A new proof of a theorem of Narasimhaus and Seshadri. Oxford University. Preprint, 1982.

[11] ———. An application of gauge theory to the topology of 4-manifolds. Oxford University. Preprint, 1982.

[12] J. Eells and L. Lemaire. A report on harmonic maps. *Bull. London Math. Soc.*, **10** (1978), 1–68.

[13] J. Eells and J. Wood. Harmonic maps from surfaces to complex projective space. University of Warwick. Preprint, 1981.

[14] R. Hamilton. Three manifolds with positive Ricci curvature. *J. Diff. Geom.*, **17** (1982), 255–306.

[15] J. Kazden and F. Warner. Scalar curvature and conformal deformation of Riemannian structure. *J. Diff. Geom.*, **10** (1975), 113–134.

[16] S. Kobayashi. Curvature and stability of vector bundles. *Proc. Japan Acad.*, **58-A** (1982), 158–162.

[17] P. Lax. Integrals of non-linear equations of evolution and solitary waves. *Comm. Pure Appl. Math.*, **21**, 647 (1968), 467–490.

[18] J. Milnor. *Morse Theory*. Annals of Math Studies, *51*. Princeton University Press: Princeton, N.J., 1970.

[19] J. Moser. *Various Aspects of Integrable Hamiltonian Systems*. Progress in Mathematics, 8. Birkhauser: Boston, 1980, pp. 233–289.

[20] E. Noether. Invariante Variationsprobleme. *Nachrichten von der Gesellschaft der Wissenschaften zu Göttingen* (1918), 235–257.

[21] R. S. Palais. *Foundations of Global Non-Linear Analysis*. Benjamin: New York, 1968.

[22] P. Price. A monotonicity formula for Yang–Mills fields. Preprint.

[23] R. Schoen and K. Uhlenbeck. A regularity theory for harmonic maps. *J. Diff. Geom.*, **17** (1982), 307–335.

[24] S. Sedlacek. A direct method for minimizing the Yang–Mills functional over 4-manifolds. *Commun. in Math. Physics*, **86** (1982), 515–528.

[25] C. H. Taubes. Self-dual connections on non-self dual 4-manifolds. *J. Diff. Geom.*, **17** (1982), 139–170.

[26] H. Yamabe. On a deformation of Riemannian structures on compact manifolds. *Osaka Math. J.*, **12** (1960), 21–37.

[27] C. N. Yang. Condition of self-duality for SO(2) gauge fields on Euclidean four-dimensional space. *Phys. Rev. Lett.*, **38** (1977), 1377–1379.

[28] S. T. Yau. On the Ricci curvature of a compact Kähler manifold and the complex Monge–Ampere equation. *Comm. Pure Appl. Math.*, **31** (1978), 339–411.

Finite Simple Groups*

WALTER FEIT†

I wish to thank the organizers of this symposium for inviting me to speak and enabling me to discuss some consequences of the classification of the finite simple groups.

The structure of finite groups is one of the few branches of algebra to which Emmy Noether made no direct contributions. It is nevertheless appropriate to talk about this subject at a symposium to celebrate the 100th anniversary of her birth, because her work has had an influence in mathematics far beyond the areas to which she made explicit contributions. The abstract and axiomatic point of view is nowadays so basic to the way we think about mathematics that we tend to forget that its origins are relatively recent. Emmy Noether was prominent among a small group of mathematicians who profoundly influenced the way mathematics is done today. Without this influence it seems to me very likely that the classification of finite simple groups would still lie in the future.

I gave a similar talk at Santa Cruz in 1979 and much of this material has appeared in [1] so that I won't repeat it here. Since that time new results have been found. For instance, I have been able to show that most trees are not Brauer trees [2]. One of the most striking applications of the classification of finite simple groups has recently been found by B. Fein, W. M. Kantor and M. Schacher [3]. Since this is a result that would have interested Emmy Noether it is very fitting to be able to state it here.

If $K \subseteq L$ are global fields let $B(L/K)$, the relative Brauer group of L/K, denote the subgroup of the Brauer group $B(K)$ consisting of those classes of finite dimensional central simple K-algebras which are split by L.

* This note contains a brief synopsis of the talk the author gave at the Emmy Noether Symposium. The work in this paper was supported by the N.S.F.

† Department of Mathematics. Yale University, P.O. Box 2155, Yale Station, New Haven, CT 06520, U.S.A.

Theorem (See [3]). *If $K \neq L$ then $B(L/K)$ is infinite.*

The proof of this theorem depends on the following:

Proposition (See [3]). *If H is a proper subgroup of the finite group G, then there exists a prime p and a p-element which is not conjugate to any element in H.*

A simple counting argument shows that $G \neq \bigcup x^{-1}Hx$, hence there exists an element in G which is not conjugate to any element of H. The proposition is an innocent looking refinement of this statement. However, the only known proof consists of reducing the question to the case that G is simple (this is not difficult) and then checking all the simple groups. The latter step, of course, uses the classification in an essential way.

References

[1] W. Feit. Some consequences of the classification of finite simple groups. *Proc. Sympos. Pure Math.*, vol. 37, Santa Cruz 1979, 405–412. AMS: Providence, 1980.

[2] ———. Possible Brauer trees. (To appear.)

[3] B. Fein, W. M. Kantor, and M. Schacher. Relative Brauer groups, II. *J. reine angew. Math.*, **328** (1981), 39–57.

L^2-Cohomology and Intersection Cohomology of Certain Arithmetic Varieties

ARMAND BOREL*

Introduction

L^2-cohomology of non-compact manifolds and intersection cohomology are two fairly recent topics. In principle, they are independent from one another, one belonging to differential geometry or analysis, the other to algebraic topology and algebraic geometry. However, some interesting relations between the two appeared almost from the beginning. To what extent they are special cases of rather general phenomena is not known at present, but, for a number of people, the evidence is tantalizing enough to make it appear worthwhile to try to investigate this further. From a broader point of view, one can view this as an attempt to add a new chapter to a topic with an already long history: the connections between differential geometry and topology or, more specifically, the representation of topological invariants by means of analytical objects.

A first well-known example is the isomorphism of the first complex cohomology group $H^1(M; \mathbf{C})$ of a compact Riemannian surface M with the direct sum $S_2(M) + \bar{S}_2(M)$ of the spaces of holomorphic and of antiholomorphic differentials. A far-reaching generalization is given by the Hodge theorem which, for a compact smooth Riemannian manifold M, provides an isomorphism between the ith cohomology group $H^i(M; \mathbf{C})$ and the space $\mathscr{H}^i(M)$ of harmonic i-forms ($i = 0, 1, \ldots$). If M is kählerian, in particular, if it is a projective variety, this can be considerably refined by the consideration of the (p, q)-types and the introduction of a Hodge structure on $H^{\cdot}(M; \mathbf{Q})$. The research alluded to above may then be viewed as an attempt to provide similar connections when M is a singular space, in particular, a projective variety. Roughly speaking, if Σ is the singular locus, one tries to establish an isomorphism between the space of L^2-harmonic forms of $M - \Sigma$ with respect to a suitable metric, and the middle intersection cohomology

* Institute for Advanced Study, School of Mathematics, Princeton, NJ 08540, U.S.A.

of M, in the sense of M. Goresky and R. MacPherson [10], [11]. So far, this has been done notably for conical singularities and conical metrics (J. Cheeger [7]), variations of Hodge structures on punctured curves (S. Zucker [14] and for the minimal compactifications of an arithmetic quotient of a bounded symmetric domain in the so-called $\mathbf{Q}$-rank one case (i.e. when the singular locus consists of one closed smooth subvariety of pure dimension). Those are the "certain arithmetic varieties" of our title. The main goal of this paper is to discuss this last case. To make the paper a bit more self-contained, we shall also recall some basic notions and facts on L^2-cohomology and on middle intersection cohomology, the latter, however, only for spaces with one singular stratum. For a more general discussion of those topics and of their proven or conjectured relationships, we refer to [8].

1. L^2-Cohomology

1.1. Let M be a Riemannian manifold. We let $A^{\cdot}(M)$ denote the space of smooth complex-valued differential forms on M. The metric defines at each point $x \in M$ a scalar product $(\ ,\ )_x$ on the exterior algebra $\Lambda T^*(M)_x$ of the cotangent space $T^*(M)_x$ at x, whence a, possibly infinite, scalar product

$$(1) \qquad (\mu, \eta) = \int_M (\mu_x, \eta_x)_x \, dv \qquad (\mu, \eta \in A^{\cdot}(M)),$$

where dv is the Riemannian volume element. We let $A^{\cdot}_2(M)$ be the space of η for which (η, η) is finite and $A^{\cdot}_{(2)}(M)$ the space of η such that $\eta, d\eta \in A^{\cdot}_2(M)$. The space $A^{\cdot}_{(2)}(M)$ is stable under the exterior differentiation d and the space $H^{\cdot}_{(2)}(M; \mathbf{C})$ of L^2-cohomology of M with complex coefficients is by definition the cohomology of $A^{\cdot}_{(2)}(M)$. Let as usual $\partial\colon A^p(M) \to A^{p-1}(M)$ be the boundary operator. It is formally adjoint to d in the sense that we have

$$(2) \qquad (\mu, \partial\eta) = (d\mu, \eta) \quad (\mu, d\mu, \eta, \partial\eta \in A^{\cdot}_2(M),\ \mu \text{ or } \eta \text{ of compact support}).$$

Let $\mathscr{H}^{\cdot}_{(2)}(M; \mathbf{C})$ be the space of L^2-harmonic forms, i.e. of the $\eta \in A^{\cdot}_2(M)$ which satisfy $d\eta = \partial\eta = 0$. There are natural homomorphisms

$$(3) \qquad \mathscr{H}^{\cdot}_{(2)}(M, \mathbf{C}) \xrightarrow{\alpha} H^{\cdot}_{(2)}(M; C) \xrightarrow{\beta} H^{\cdot}(M; \mathbf{C}),$$

which are isomorphisms if M is compact, but are neither surjective nor injective in general. If M is complete, our main case of interest in this paper, then α is injective (see, for example, [2]) and (2) is valid for $\mu, d\mu, \eta, \partial\eta \in A^{\cdot}_2(M)$ (a result due originally to M. Gaffney, see, for example, [2] for a proof).

1.2. The cokernel of α is either zero or infinite dimensional. Thus if we wish to compare L^2-cohomology with some topological invariant which is known to be finite dimensional, a prerequisite is that $\mathscr{H}^{\cdot}_{(2)}(M; \mathbf{C})$ be equal to $H^{\cdot}_{(2)}(M; \mathbf{C})$ and be finite dimensional. One may then wonder why one should look at $H^{\cdot}_{(2)}$ at all, rather than limit oneself to $\mathscr{H}^{\cdot}_{(2)}$. The point is that, in order to try to set up such isomorphisms, one often goes over to sheaves of germs of forms. This can be done easily enough with $A^{\cdot}_{(2)}(M)$ (see below for some illustrations), but hardly with harmonic forms.

1.3. A theorem of K. Kodaira asserts that β and $\beta \circ \alpha$ have the same image. When $M = \Gamma\backslash X$ is a non-compact quotient of a negatively curved symmetric space X by an arithmetic group, for instance when $N = \mathbf{SL}_n(\mathbf{Z})\backslash\mathbf{SL}_n(\mathbf{R})/\mathbf{SO}(n)$, it has proved very useful to show that $\beta \circ \alpha$ is an isomorphism in some range, which can be described in terms of roots [2], [3]. More recently, S. Zucker has shown the existence of such a range in which β itself is an isomorphism [15].

1.4. Let E be a local system on M defined by a finite dimensional unitary representation (τ, E_0) of the fundamental group of M. There is then a scalar product on $\Lambda T^*(M)_x \otimes E_x$ and the formula 1.2(1) again defines a scalar product on the space $(M; E)$ of E-valued smooth differential forms. We let $A^{\cdot}_{(2)}(M; E)$ be the complex $A^{\cdot}$ of smooth E-valued forms η such that η and $d\eta$ are square integrable. Its cohomology is the space $H^{\cdot}_{(2)}(M; E)$ of L^2-cohomology of M with coefficients in E. The previous remarks extend obviously to this case.

1.5. We refer to [7] for Hilbert space definitions of L^2-cohomology.

2. Middle Intersection Cohomology for Spaces with a Smooth Singular Set

2.1. Let V be a locally compact space and S a closed subspace. We assume that $V - S$ is a smooth orientable n-manifold and that S is a manifold of pure *even* codimension $2m$, the "singular stratum" of V. Moreover, for $x \in S$, it is assumed that there exists a closed submanifold L_x of $V - S$, called the link of x, such that x has in V a fundamental set of neighborhoods U homeomorphic to a product of a neighborhood B of x in S by a cone $c(L_x)$ over L_x with vertex x. There are some further regularity assumptions which we do not spell out, such as the existence of a PL-structure on V, L_x for which S is a subcomplex and $U \tilde{\to} B \times c(L_x)$ is PL, etc. cf. [10], [11].

2.2. Let k be a field $C. = C.(V; k)$ be the inductive limit of the simplicial chains with coefficients in k for triangulations compatible with the given PL structure (the chains are allowed to be infinite if V is not compact). Then $IC.(V; k)$ is the space of i-chains ξ satisfying

$$(1) \qquad \dim(\xi \cap S) \leq i - m - 1, \qquad \dim(\partial\xi \cap S) \leq i - 1 - m - 1,$$

and the ith *middle intersection homology space* is

$$(2) \qquad IH_i(V; k) = H_i(IC.(V; k)).$$

To understand this, one should write the $-m - 1$ on the right-hand sides in (1) as $-2m + (m - 1)$. The difference $i - 2m$ would be the dimension of $\xi \cap S$ if ξ would intersect S in general position. If we replace $m - 1$ by 0 then, by using dual cell subdivisions, one sees that one gets a complex defining cohomology (in complementary dimensions), at any rate if the links are connected [10]. If we replace $m - 1$ by $2m$, then there is no condition at all and we get homology. In fact, we already do so if we replace $m - 1$ by $2m - 2$ and assume again the links to be connected [10]. The $m - 1$ is the "perversity." It could be replaced by any integer between 0 and $2m - 2$ and this would lead to the definition of other intersection homology groups. By choosing $m - 1$, we are in the middle of the possible perversities (or also in some sense half-way between homology and cohomology), whence the expression "middle intersection homology." The homology groups $IH_i(V; k)$ are topological invariants and, moreover, finite dimensional if V is compact.

2.3. If $i \leq m$, the right-hand side of the first relation in 2.2(2) is ≤ -1, hence ξ must have its support in $V - S$. If $i > n - m$, then the right-hand side is $\geq n - 2m$, which is the dimension of S, hence the condition imposed on ξ itself is vacuous. From this and similar remarks on the second condition in 2.2(1) we see that if V is compact, then

(1) $$IH_i(V; k) = H_i(V; k) \quad \text{if } i > n - m,$$

(2) $$IH_i(V; k) = H_i(V - S) \quad \text{if } i \leq m - 1,$$

(3) $$IH_m(V; k) \text{ is a quotient of } H_m(V - S; k),$$

(4) $$IH_{n-m}(V; k) \text{ is a subspace of } H_{n-m}(V; k).$$

If now $n = 2m$, and $i = m$, then the boundaries in IC_i are those of arbitrary chains while a cycle in IC_m is any cycle with support in $V - S$. As a consequence

(5) $$IH_m(V; k) = \operatorname{Im}((H_m(V - S; k) \to H_m(V; k)) \quad \text{if } n = 2m.$$

This is also true if V is not compact, provided ordinary homology is understood with infinite chains.

2.4. One can also define similarly intersection homology with respect to a local system E of finite dimensional vector spaces over k on $V - S$. Note that any simplex in $IC.$ is not contained in S, and this allows one easily to define the complex $IC.(V; E)$ of intersection chains with coefficients in E, and the corresponding middle intersection homology groups $IH.(V; E)$. By taking finite chains one defines intersection homology with compact supports.

2.5. It will be more convenient for us to use a cohomological grading. We set, by definition

(1) $$IC^i(V; E) = IC_{n-i}(V; E), \qquad IH^i(V; E) = IH_{n-i}(V; E) \qquad (i \in \mathbf{N}),$$

and similarly for compact supports. If S is empty, this means that we take Poincaré duality, when stated as an isomorphism between homology and cohomology, as a definition! However, Poincaré duality becomes again a theorem when formulated as a pairing between cohomology groups of complementary dimensions.

2.6. *Poincaré duality*. One main property of the middle intersection cohomology is that it satisfies Poincaré duality: There exists a perfect pairing

$$IH_c^i(V; E) \times IH^{n-i}(V; E^*) \to k \qquad (i \in \mathbf{N}),$$

where E^* is the local system contragredient to E.

2.7. *Local groups.* By using locally finite simplicial chains, we can define a sheaf $\mathscr{I}\mathscr{C}^*(V; E)$ of germs of middle intersection cochains. It is fine. The stalks of the derived sheaf $\mathscr{H}(\mathscr{I}\mathscr{C}^\cdot)$ are given by

(1) $$\mathscr{H}^q(\mathscr{I}\mathscr{C}^\cdot)_x = \begin{cases} E & \text{if } q = 0, \\ 0 & \text{if } q \geq 1. \end{cases} \qquad (x \in V - S);$$

(2) $$\mathscr{H}^q(\mathscr{I}\mathscr{C}^\cdot)_x = \begin{cases} H^q(L_x; E) & \text{if } q < m, \\ 0 & \text{if } q \geq m. \end{cases} \qquad (x \in S)$$

For $x \in S$, let

$$H^q_x(\mathscr{I}\mathscr{C}^\cdot) = \varprojlim_U IH^q_c(U; E) \qquad (q \in \mathbf{N}),$$

where U runs through a fundamental set of neighborhoods of x. Note the place of the subscript x. A more orthodox notation for the left-hand side would be $H^q(f_x^!\mathscr{I}\mathscr{C}^\cdot)$, where f_x is the inclusion of x in X. We also have

(3) $$\mathscr{H}^q_x(\mathscr{I}\mathscr{C}^\cdot) = \begin{cases} 0 & (q \leq n - m), \\ H^{2m-1-n+q}(L_x; E) & (q > n - m). \end{cases}$$

2.8. These properties essentially characterize the intersection cohomology sheaf, up to quasi-isomorphism. To be more precise let $\mathscr{S}$ be a differential graded sheaf on V. Assume

(i) $\mathscr{S}$ is fine, graded by positive degree $i\mathscr{S}|_{V-S}$ is quasi-isomorphic to the locally constant sheaf $\mathscr{E}$ defined by E. The stalks of $\mathscr{H}\mathscr{S}$ on S form a locally constant sheaf on S.
(ii) The stalks of $\mathscr{H}\mathscr{S}$ satisfy 2.7(1) and either

(1) 2.7(2), *where the isomorphisms for $q < m$ satisfy a certain naturality condition, the "attachment condition,"*

or

(2) $\mathscr{H}^q(\mathscr{S})_x = 0$ for $q \geq m$ and $\mathscr{H}^q_x(\mathscr{S}) = 0$ for $q \leq n - m$,

then $\mathscr{S}$ is quasi-isomorphic to $\mathscr{I}\mathscr{C}^\cdot(V; E)$. In particular

(3) $$H^\cdot(\Gamma(\mathscr{S})) = IH^\cdot(V; E), \qquad IH^\cdot(\Gamma_c(\mathscr{S})) = IH^\cdot_c(V; E),$$

where Γ and Γ_c refer to continuous and compactly supported continuous sections, respectively. Here $\mathscr{H}_x(\mathscr{S})$ is defined by

(4) $$\mathscr{H}_x(\mathscr{S}) = \varprojlim_U H^i(\Gamma_c(\mathscr{S})).$$

In fact, this characterization holds also for $\mathscr{S}$ not fine, but one has then to use hypercohomology.

2.9. We have assumed $V - S$ to be orientable for convenience. If not, one has to twist by the orientation sheaf at appropriate places. For the proofs of all of the above statements, in much greater generality, see [11].

3. A Method to Relate L^2-Cohomology and Intersection Cohomology

3.1. We keep the setup of Section 2. Assume we are given a Riemannian metric on $M = V - S$ and are interested in knowing whether $H^{\cdot}_{(2)}(M; E) = IH^{\cdot}(V; E)$. We describe one way to try to go about it.

Consider on V the presheaf which associates to an open subset U the complex $A^{\cdot}_{(2)}(M \cap U; E)$. This presheaf is in fact a sheaf $\mathscr{A}^{\cdot}$, and the space of its continuous sections over V is just $A^{\cdot}_{(2)}(M; E)$. *Assume it is fine*. Then, in order to prove that $H^{\cdot}_{(2)}(M; E) = IH^{\cdot}(V; E)$, it suffices to show that $\mathscr{A}^{\cdot}$ is quasi-isomorphic to $\mathscr{I}\mathscr{C}^*$ and for this it is enough to see that the stalks of the derived sheaf $\mathscr{H}\mathscr{A}^{\cdot}$ satisfy one of the sets of requirements in 2.8. The conditions 2.8(i) and 2.7(1) are obvious, so what remains is to study the derived sheaf $\mathscr{H}\mathscr{A}^{\cdot}$ at points of S. By construction, $H^{\cdot}(\Gamma(\mathscr{A} \mid U)) = H^{\cdot}_{(2)}(U \cap M; E)$, therefore

(1) $$\mathscr{H}^{\cdot}(\mathscr{A}^{\cdot})_x = \varinjlim_U H^{\cdot}_{(2)}(U \cap M; E),$$

where U runs through a fundamental set of neighborhoods of x in V. The space $\Gamma_c(\mathscr{A} \mid U)$ may be identified with the space $A^{\cdot}_{(2),rc}(U \cap M; E)$ of elements in $A^{\cdot}_{(2)}(U \cap M; E)$ whose support is relatively compact in V. We have therefore

(2) $$\mathscr{H}^{\cdot}_x(\mathscr{A}^{\cdot}) = \varprojlim_U H^{\cdot}(A^{\cdot}_{(2),rc}(U \cap M; E)).$$

In practice, these inductive or projective systems are essentially constant, at any rate for a suitable fundamental set of neighborhoods. Therefore the main burden is to prove either that $H^{\cdot}_{(2)}(U \cap M; E)$ satisfies 2.7(2) or that

(3) $$H^q_{(2)}(U \cap M; E) = 0 \quad \text{for } q \geq m \quad \text{and}$$
$$H^r(A_{(2),rc}(U \cap M; E)) = 0 \quad \text{for } r \leq n - m.$$

The impression I have gained so far is that the crucial test is whether the first part of (3) can be proved. If so, it is not unlikely that the argument will also yield 2.7(2) for $i < m$ and that some dual version of it will give the second part of (3). Each of these two suffices to complete the proof of the isomorphism.

3.2. Cheeger [7] considers (not exclusively) the case where $n = 2m$, i.e. where S consists of isolated points and the metric around $x \in S$ is "conical": In this case $U = c(L)$ is the cone over the link. If r is a coordinate along the generating lines of the cone, with x at the origin, then the metric is of the form $ds^2 = dr^2 + r^2\, ds_L^2$. It is not complete. The gist of the matter is indeed the computation of the L^2-cohomology of $c(L) - \{x\}$.

3.3. In [4], V is a complete smooth complex curve and the metric around $x \in S$ is the Poincaré metric of a punctured disc. If we view a neighborhood of x as a cone $c(L)$ over a circle L with coordinate t, the metric around x can be written

$$ds^2 = r^{-2} \cdot dr^2 + e^{-2/r} \cdot dt^2.$$

It is complete. If now E represents a variation of Hodge structure on M and $j: M \hookrightarrow V$ is the inclusion, then [14] proves an isomorphism

$$(1) \qquad H^{\cdot}(V; j_* E) = H^{\cdot}_{(2)}(M; E).$$

It was seen later that the left-hand side is indeed middle intersection cohomology, so that this result could also be viewed as another example of an isomorphism between the latter and L^2-cohomology.

4. Locally Symmetric Varieties

4.1. In the sequel, $\mathscr{G}$ is a semi-simple connected **Q**-group, almost simple over **Q**. Let $G = \mathscr{G}(\mathbf{R})$ be the group of real points of $\mathscr{G}$ and K a maximal compact subgroup of G. We assume that $X = G/K$ is a bounded symmetric domain. Let Γ be an arithmetic subgroup of G. Then $M = \Gamma\backslash X$ has a canonical structure of quasi-projective variety and a minimal projective compactification [1], to be denoted V. To simplify we assume that Γ is neat (the subgroup of $\mathbf{C}^*$ generated by the eigenvalue of any element of Γ is torsion-free). In particular, Γ is torsion-free, hence $M = \Gamma\backslash X$ is a complex manifold. We assume, moreover, that G has **Q**-rank one (the maximal **Q**-split tori of $\mathscr{G}$ have dimension one). This implies first that M is not compact, hence $M \neq V$ and second that $S = V - M$ is itself a smooth projective variety. In fact, its connected components are smooth compact arithmetic quotients of lower-dimensional bounded symmetric domains, the "rational boundary components" [1].

4.2. We give here some examples.

(1) Let F be a totally real field, d its degree over **Q** and $\mathfrak{o}_F$ the ring of integers of F. Take for G the group $R_{k/\mathbf{Q}}\mathbf{SL}_2$ obtained by restriction of scalars from $\mathbf{SL}_2$ viewed as a k-group. Then $G = \mathbf{SL}_2(\mathbf{R})^d$, and X is the product H^d of d copies of the upper halfplane H. Take for Γ a subgroup of finite index of $\mathbf{SL}_2\,\mathfrak{o}_F$. Then $\Gamma\backslash X$ is a Hilbert–Blumenthal variety and V is obtained by compactifying M with finitely many points.

(2) Let k be a quadratic imaginary field, h a hermitian form over k^n of Witt index one and of signature (p, q) over **C** $(p \geq q \geq 1)$. This defines a **Q**-form of $\mathbf{SL}_n$ such that $G = \mathbf{SU}(p, q)$. In that case, X may be identified with an open subset of the Grassmannian $\mathbf{G}_{p,n}$ of p planes in $\mathbf{C}^n$ and, similarly, S with an open subset of $\mathbf{G}_{p-1,n-2}$. For $q = 1$, X is the unit ball in $\mathbf{C}^p$. In this case, Γ is a subgroup of finite index of the group of units of h.

(3) The group G is a **Q**-form of **Q**-rank one the exceptional group $\mathbf{E}_6$ of type ${}^2\mathbf{E}^{16}_{6,2}$ over **R** in the notation of [12]. Then X is 16-dimensional and each connected component of S is a compact quotient of the unit ball in $\mathbf{C}^5$.

4.3. The space M has a natural Riemannian metric, stemming from a G-invariant metric on X or from the Bergmann metric in any realization of X as a bounded domain. It is complete.

On M we consider a local system E associated to a Γ-module which is an irreducible rational representation E_0 of $\mathscr{G}$. Then E is not unitary with respect to Γ, unless it is trivial, so that 1.4 is not directly applicable. However, it is well-known and easily checked that $A^{\cdot}(M; E)$ can also be viewed as the space of smooth sections of a K-bundle over $\Gamma\backslash X$, with typical fiber $\Lambda T^*(M)_0 \otimes E_0$, whence the existence of a scalar product. Hence $H^{\cdot}_{(2)}(M; E)$ is again well-defined.

4.4. Theorem. *We keep the notation and assumptions of* 4.1. *Then* $H^{\cdot}_{(2)}(M; E)$ *is isomorphic to* $IH^{\cdot}(V; E)$.

In [15], S. Zucker conjectures that this theorem is true without any restriction on the **Q**-rank and also checks it in some cases where the **Q**-rank is one, in particular for Hilbert–Blumenthal varieties.

The validity of this conjecture in that last case has been used by Brylinski and Labesse [6] to express the Hasse–Weil zeta function of a canonical model of V, with respect to middle intersection cohomology, or rather to its analogue in the étale cohomology setting, in terms of automorphic L-functions. In fact, a prime reason of interest in this conjecture is the hope that, if true, it will help to extend such results to other arithmetic quotients of bounded symmetric domains.

5. Sketch of the Proof of Theorem 4.4.

5.1. For $v \in \mathbf{R}$ and f a function on the half line $\mathbf{R}^+ = [0, \infty)$ we define a "v-norm" $\|\cdot\|_v$ by:

$$\|f\|_v^2 = \int_0^\infty |f(r)|^2 e^{2vr}\, dr. \tag{1}$$

By $H^{\cdot}(\mathbf{R}^+, v)$ we denote the cohomology of the complex $C^{\cdot}_v$, where C^0_v consists of the smooth functions such that f, df/dr have finite v-norms and C^1_v of the one-forms $f \cdot dr$, where f is smooth with finite v-norm. Then, as a special case of 3.69 in [15], we have:

$$H^0(\mathbf{R}^+, v) = \begin{cases} 0 & \text{if } v \geq 0, \\ \mathbf{C} & \text{if } v < 0, \end{cases} \qquad \dim H^1(\mathbf{R}^+, v) = \begin{cases} 0 & \text{if } v \neq 0, \\ \infty & \text{if } v = 0. \end{cases} \tag{2}$$

Let now $C^{\cdot}_{v,c}$ be the complex defined as above, except that we consider only functions which vanish in a neighborhood of the origin. Let $H^{\cdot}_c(\mathbf{R}^+, v)$ be the cohomology of $C^{\cdot}_{v,c}$. Then we have obviously

$$H^0_c(\mathbf{R}^+, v) = 0; \tag{3}$$

a computation similar to the one yielding (2) shows:

$$\dim H^1_c(\mathbf{R}^+, v) = \begin{cases} 0 & \text{if } v < 0, \\ \infty & \text{if } v = 0, \\ 1 & \text{if } v > 0. \end{cases} \tag{4}$$

These results remain true if $\mathbf{R}^+$ is replaced by $R^{\geq t} = [t, \infty)$ and in fact the maps

$$H^\cdot(\mathbf{R}^+, v) \to H^\cdot(\mathbf{R}^{\geq t}, v) \quad \text{and} \quad H_c(\mathbf{R}^{\geq t}, v) \to H_c(\mathbf{R}^+, v),$$

defined by restriction and inclusion are isomorphisms ($t > 0$).

5.2. We now go on in the setting of Section 4 and want to use the approach outlined in Section 3. We can start because the L^2-sheaf $\mathscr{A}$ on V is indeed fine: in [15], this is proved for certain Satake compactifications but the argument applies to all, and V is one. We have then to consider the L^2-cohomology around a point $x \in S$. A typical neighborhood U of x in V can be written $U = B \times c(L)$, where B is a ball in S with center x and L the link of x. Then

(1) $$U \cap M = B \times c(L)^0, \quad \text{where} \quad c(L)^0 = c(L) - \{x\}.$$

The metric on $U \times M$ is the product metric and a Künneth rule [15: 2.36] allows one to discard U. We are therefore reduced to considering $H^\cdot_{(2)}(c(L)^0; E)$. We shall write $c(L)^0$ as

(2) $$c(L)^0 = \mathbf{R}^+ \times L,$$

with the understanding that x is at infinity. With respect to the metric, this is an orthogonal decomposition but the space is by far not a metric product: The metric on $\{r\} \times L$ depends on r in a rather complicated way. It is warped [15].

The space L is fibered over a compact quotient Y of a symmetric space (not complex in general), with fibre a compact nilmanifold. To be more precise, let P be the group of real points of a minimal parabolic $\mathbf{Q}$-subgroup $\mathscr{P}$ of $\mathscr{G}$ and let $P = N.A.{}^0M$ be its usual Langlands decomposition: N is the unipotent radical of P and ${}^0M.A$ a Levi subgroup. Under the projection $\sigma: P \to P/N \simeq {}^0M.A$, the group A maps onto the identity component of the group of real points of a maximal (one-dimensional) $\mathbf{Q}$-split torus of P/N and ${}^0M$ is defined as the intersection of the kernels of certain characters. We have $Z_G(A) = {}^0M.A = {}^0M \times A$. Let $\Gamma_P = \Gamma \cap P$ and $\Gamma_N = \Gamma \cap N$. Then $\Gamma_P \subset {}^0M.N$. We assume ${}^0M$ and A to be stable under the Cartan involution θ associated to K. The rational boundary components correspond to the Γ-conjugacy classes of minimal parabolic $\mathbf{Q}$-subgroups of G. We assume P to be associated to the boundary component S_0 containing x. There exists a decomposition ${}^0M^0 = H.Z$ of the identity component of ${}^0M$ into an almost direct product of two normal $\mathbf{Q}$-subgroups H, Z, also stable under θ, with H semi-simple, such that the symmetric space $X_H = H/K_H$ ($K_H = K \cap H$)) is a bounded domain and the universal covering of S_0 and that the link L admits $N.Z/K_Z$ ($K_Z = K \cap Z$) as a universal covering. Let

$$\Gamma_1 = \Gamma \cap Z.A.N., \qquad \Gamma_Z = \Gamma_1/\Gamma_N, \qquad \Gamma_H = \Gamma_P/\Gamma_1.$$

Then $\Gamma_1 \subset Z.N$ and we have:

$$S_0 = \Gamma_H \backslash X_H \quad \text{and} \quad L = \Gamma_1 \backslash N.Z/K_Z.$$

This shows in particular that L is fibered over $Y = \Gamma_Z \backslash X_Z$, with typical fiber $\Gamma_N \backslash N$. Note that Γ_Z and Γ_H are torsion free since Γ is assumed to be neat.

We identify $\mathbf{R}$ to the Lie algebra of A and therefore A to the multiplicative group of strictly positive numbers. Let $A^+ = \exp \mathbf{R}^+ = \{r \in \mathbf{R}, r \geq 1\}$. We view $U \cap M$ as a fiber bundle over $Y \times A^+ = Y \times \mathbf{R}^+$, with typical fiber $\Gamma_N \backslash N$.

5.3. By a result of van Est [13] (proved first by K. Nomizu for trivial coefficients), we have

$$H^{\cdot}(\Gamma_N \backslash N; E) = H^{\cdot}(\mathfrak{n}; E), \tag{1}$$

where the right-hand side refers to Lie algebra cohomology, $\mathfrak{n}$ being the Lie algebra of N.

In ordinary cohomology, it is known that the spectral sequence of the fibration of L over Y degenerates at E_2, therefore

$$H^{\cdot}(L; E) = H^{\cdot}(Y; H^{\cdot}(\mathfrak{n}; E)) = H^{\cdot}(Y \times \mathbf{R}^+; H^{\cdot}(\mathfrak{n}; E) = H^{\cdot}(c(L)^0; E). \tag{2}$$

Note that, since X_Z is contractible to a point, we have

$$H^{\cdot}(Y; H^*(\mathfrak{n}; E)) = H^{\cdot}(\Gamma_Z; H^{\cdot}(\mathfrak{n}; E)), \tag{3}$$

where the right-hand side denotes Eilenberg–MacLane cohomology. In the L^2-case, there is a partial analogue, namely:

$$H^{\cdot}_{(2)}(c(L)^0; E) = H_{(2)}(Y \times \mathbf{R}^+; H^{\cdot}(\mathfrak{n}; E)). \tag{4}$$

However, the factor $\mathbf{R}^+$, which can be ignored in the topological case, plays an important role in the L^2-setting.

According to a theorem of B. Kostant (recalled in [5: III] or also in [15]) there is a multiplicity-free decomposition of $H^*(\mathfrak{n}; E)$ into irreducible rational ${}^0M.A$ modules (also irreducible under the identity component ${}^0M^0A$ of ${}^0M.A$). Write then

$$H^i(\mathfrak{n}; E) = \bigoplus_{1 \leq j \leq n_i} F_{i,j} \qquad (i \in \mathbf{N}), \tag{5}$$

where $F_{i,j}$ is such a module. Then $F_{i,j}$ is the tensor product of an irreducible ${}^0M$-module by a one-dimensional A-module. Let $v(i, j)$ be the real number such that e^r acts on $F_{i,j}$ by $e^{v(i,j)r}$. The $\gamma(i, j)$'s and, more precisely, the highest weights of the $F_{i,j}$ are explicitly described by Kostant's theorem. This information plays a basic role in the proof. However, since we leave out the computations, we do not need it explicitly here and omit it. Let $\rho \in \mathbf{R}$ be defined by:

$$e^{2\rho.r} = \det \operatorname{Ad} e^r | \mathfrak{n} \qquad (r \in \mathbf{R}). \tag{6}$$

Then we have

$$H^{\cdot}(c(L)^0; E) = \bigoplus_{i,j} H^{\cdot}(\mathbf{R}^+, v(i,j) - \rho) \otimes (H^{\cdot}(\Gamma_Z; F_{i,j})[-i]). \tag{7}$$

(As usual, if $C^{\cdot}$ is a graded complex and $i \in \mathbf{Z}$, then $C^{\cdot}[i]$ is the graded complex defined by $(C^{\cdot}[i])^j = C^{j+i}$ $(j \in \mathbf{Z})$.) This equality, in a different notation, is contained in 3,vi of [15].

On the other hand, we can write, by (2), (3), (5),

(8) $$H^\cdot(L; E) = \bigoplus_{i,j} H^\cdot(\Gamma_Z; F_{i,j})[-i],$$

or, more explicitly,

(9) $$H^q(L; E) = \bigoplus_{i,j} H^{q-i}(\Gamma_Z; F_{i,j}) \qquad (q \in \mathbf{N}).$$

5.4. We get a fundamental system of neighborhoods $\{U_t\}$ of x just by replacing in the above $\mathbf{R}^+$ by $\mathbf{R}^{\geq t}$ (t tending to infinity). This does not change the ordinary cohomology of $U \cap M$ (obviously) or the L^2-cohomology and the groups $H^\cdot(A^\cdot_{(2),rc}(U \cap M; E))$ of 3.1 by the end remark of 5.1. We have therefore to check the conditions of 2.8 directly for $U \cap M$.

The first step is to make sure that no infinite dimensional cohomology occurs. In view of 5.1(2) and 5.3(7), this amounts to proving that $H^\cdot(\Gamma_Z; F_{i,j}) = 0$ whenever $v(i,j) = \rho$. This can be shown *a priori* from the results of Section 1 in [4]. We assume this has been done. Then we can write

(10) $$H^q(c(L)^0; E) = \bigoplus_{i,j} H^0(\mathbf{R}^+, v(i,j) - \rho) \otimes H^{q-i}(\Gamma_Z; F_{i,j}) \qquad (q \in \mathbf{N}).$$

In view of 5.1, the vanishing part of 2.7(2) is equivalent to:

(*) *Let* $q \geq m, i, j \in \mathbf{N}$ *and* $j \leq n_i$. *If* $v(i,j) < \rho$, *then* $H^{q-i}(\Gamma_Z; F_{i,j}) = 0$.

Note that, apart from the attachment condition, which causes no difficulty here, 2.7(2) for $q < m$ is equivalent to

(**) *Let* $q < m$, *and* $i \in \mathbf{N}$. *If* $H^{q-i}(\Gamma_Z; F_{i,j}) \neq 0$, *then* $v(i,j) < \rho$.

We concentrate here on (*). It is first rather easily seen that the proof can be reduced to the case where G is absolutely almost simple; then X is an irreducible bounded symmetric domain. From that point on, however, I have to proceed case by case using J. Tits' classification [12].

A fairly simple case is when Y is a point. Then the formulas (9), (10) simplify to

(11) $$H^q(L; E) = \bigoplus_{1 \leq j \leq n_q} F_{q,j};$$

(12) $$H^q_{(2)}(c(L)^0; E) = \bigoplus_{1 \leq j \leq n_q} H^0(\mathbf{R}^+; v(q,j) - \rho) \otimes F_{q,j}.$$

In this case (*) follows rather simply from some root and Weyl group considerations and the explicit description of the highest weight of the $F_{i,j}$'s.

There are however some cases where Y is not a point. Some information about the cohomology of Γ_Z is then needed. Since Γ_Z is cocompact, the space $L^2(\Gamma_Z \backslash Z)$ is a countable Hilbert direct sum of irreducible unitary Z-modules, say

(13) $$L^2(\Gamma_Z \backslash Z) = \widetilde{\bigoplus_{i \in I}} H_i,$$

and we have, for a finite dimensional Z-module F:

(14) $$H^{\cdot}(\Gamma_Z; F) = \bigoplus_{i \in I} H^{\cdot}(\mathfrak{z}, K_Z; H_i^\infty \otimes F).$$

(See [5: VII, 5.2]; the right-hand side refers to relative Lie algebra cohomology. Only finitely many summands are non-zero.)

Given a finite dimensional Z-module F, let $c(Z, F)$ be the greatest integer q such that there exists some irreducible unitary Z-module H for which

$$H^q(\mathfrak{z}, K_Z; H^\infty \otimes F) \neq 0.$$

For instance, if $Z = \mathbf{SL}_2(\mathbf{R})$, then $c(Z, \mathbf{C}) = 2$ and $c(Z, F) = 1$ for every irreducible non-trivial F. Now if $q - i > c(Z, F)$ then $H^{q-i}(\Gamma_Z; F) = 0$. Therefore (*) is implied by

(***) *Let $q \geq m$ and $i, j \in \mathbf{N}$, with $j \leq n_i$; assume that $v(i, j) < \rho$. Then $q > c(Z, F_{i,j}) + i$.*

This condition can be verified in the remaining cases. The most interesting is the one labeled ${}^2\mathbf{A}^1_{n,d}$ in [12]. In this case $d/n + 1$ and, over $\mathbf{Q}$,

$$G = \mathbf{SU}_{(n+1)/d}(D, h),$$

where D is a central division algebra of the second kind over a quadratic imaginary field and h a non-degenerate hermitian form of index one on $D^{(n+1)/d}$. Then, $Z = \mathbf{SL}_d(\mathbf{C})$ and the necessary information on $c(Z, F_{i,j})$ is contained in Enright's paper [9].

References

[1] W. Baily and A. Borel. Compactification of arithmetic quotients of bounded symmetric domains. *Ann. Math.*, **84** (1966), 442–528.

[2] A. Borel. Stable real cohomology of arithmetic groups. *Ann. E.N.S., Paris* (4), **7** (1974), 235–272.

[3] ———. Stable real cohomology of arithmetic groups, II. *Manifolds and Lie Groups*. Papers in honor of Y. Matsushima. Progress in Math., 14. Birkhäuser: Boston, 1981, pp. 21–55.

[4] ——— and W. Casselman. L^2-cohomology of locally symmetric manifolds of finite volume. (To appear).

[5] ——— and N. Wallach. *Continuous Cohomology, Discrete Subgroups and Representations of Reductive Groups*. Annals of Math. Studies, 94. Princeton University Press: Princeton, NJ, 1980.

[6] J. L. Brylinski and J. P. Labesse. Cohomologie d'intersection et fonctions L de certaines variétés de Shimura. Preprint.

[7] J. Cheeger. On the Hodge theory of Riemannian pseudomanifolds. *Proc. Symp. Pure Math.*, 36. R.M.S.: Providence, R.I., 1980, pp. 91–146.

[8] ———, M. Goresky, and R. MacPherson. L^2-cohomology and intersection homology. In *Seminar on Differential Geometry*, edited by S. T. Yau. Annals of Math. Studies, 102. Princeton University Press: Princeton, N.J., 1982, pp. 303–340.

[9] T. Enright. Relative Lie algebra cohomology and unitary representations of complex Lie groups. *Duke Math. J.*, **46** (1979), 513–525.

[10] M. Goresky and R. MacPherson. Intersection homology theory. *Topology*, **19** (1980), 135–162.

[11] ——— Intersection homology theory, II. (To appear in *Invent. Math.*)

[12] J. Tits. Classification of algebraic semisimple groups. *Proc. Symp. Pure Math.*, 9. A.M.S.: Providence, R.I., 1966, pp. 33–62.

[13] W. T. van Est. A generalization of the Cartan Leray spectral sequence, I, II. *Proc. Koninkl. Ned. Ad. v. Wet.-Amsterdam, Series A*, **61** (1958), 399–413.

[14] S. Zucker. Hodge theory with degenerating coefficients: L^2-cohomology in the Poincaré metric. *Ann. Math.*, **109** (1979), 415–476.

[15] ———. L^2-cohomology of warped products and arithmetic groups. *Invent. Math.* **70** (1982), 169–218.

Emmy Noether in Erlangen and Göttingen*

EMILIANA P. and GOTTFRIED E. NOETHER†

The Noether ancestors were well-to-do Jewish tradesmen engaged in the iron business in the Black Forest area of southern Germany. In the early nineteenth century, two brothers, Joseph and Hermann, moved to Mannheim and founded a wholesale iron business that remained in Noether hands until the takeover by the Nazis. In 1838, Hermann married the daughter of a Mannheim businessman, whose side interest seems to have been the study of mathematics. Hermann and his wife had three sons and two daughters. Two sons continued in the family business. But the third son, Max, born in 1844 had no inclination for a business career. He had the ill-fortune to be stricken with polio at the age of 14 and remained handicapped for the remainder of his life. He studied astronomy at the observatory in Mannheim and later mathematics at the Universities of Heidelberg, Giessen, and Göttingen. In 1868, Heidelberg awarded him the Ph.D. without his having written a dissertation. After 5 years as a Privatdozent in Heidelberg, he went to Erlangen, where he held a chair in mathematics until his death in 1921.

In 1880, Max Noether married Ida Kaufmann who came from a wealthy Jewish family in Cologne. Emmy, born in 1882, was the eldest of four children. Three brothers followed. Of the four children, Emmy and my father Fritz followed in their father's footsteps and became mathematicians.

The Noether family belonged to the intellectual middle class in Erlangen, and their daily life was marked by orderly customs and great stability. It respected education for its own sake and was interested in intellectual pursuits. Home life must have been warm and companionable, qualities that were to be associated with Emmy's own character throughout her life.

As a child Emmy gave no sign of precociousness or extraordinary ability and was undistinguishable from all the other young girls in Erlangen. From 1889 to

* Prepared by both authors for the Panel Discussion "Emmy Noether in Erlangen, Göttingen, and Bryn Mawr," March 18, 1982, and read by Gottfried E. Noether.

† The University of Connecticut, Storrs, CT 06268, U.S.A.

1897 she attended the Höhere Töchter Schule in Erlangen, at which, with other daughters of the bourgeoisie, she studied German and arithmetic, became proficient in French and English, and learned to play the piano. As she grew to young womanhood she developed a love of dancing and used to look forward to family parties. Her schoolmates and friends of this period remembered her as a clever, friendly, and rather endearing child. In 1900, at the age of 18, she took the Bavarian state examinations to become certified as a teacher of English and French. The examinations lasted several days, were rather demanding and strenuous, but she did well in them. It would seem that now her education was completed. She had gone through the required schooling deemed necessary for a young woman of her social class and breeding. Moreover, she had acquired a certificate that would enable her to earn her livelihood, should the need ever arise.

It is at this point that Emmy Noether abandoned the path followed by middle-class young women in Germany, or elsewhere in Europe for that matter, before the First World War. For it was in 1900 that she decided to attend university. It was not easy for a woman to do so in Germany in the early years of this century. Women were allowed to audit courses, but only with the professor's permission; they were not permitted to take examinations, except again by special fiat from the instructor. Germany was one of the last countries to allow women to matriculate at its universities. France had done so in 1861, England in 1878, and Italy in 1885.

In Germany conservatives held fast against any change in the traditional position of women either legally or socially, and academia reflected this attitude. In 1895 a Berlin newspaper reported that the historian Heinrich von Treitschke and a colleague at the University of Berlin had asked women to leave their classes. Subsequently, both men denied having done so, but the incident, whether true or not, impelled another professor to query academicians throughout Germany on how they felt about admitting women as regular students at the university, and what they thought of women's mental ability to do the work required. The answers make interesting reading.

Scholars in traditional fields of study, like humanities and theology, were on the whole negative to the idea of admitting women on an equal footing with men. Scientists and mathematicians were more broad-minded. But the majority of the respondents frowned upon higher education for women as being both unwomanly and beyond their mental abilities. The role of German women in life was clearly defined. They were to be wives and mothers, and any education beyond basics and a few frills like music, art, and modern languages was irrelevant and superfluous.

Many sincerely believed that the presence of women in academia would undermine its very foundations and change the nature of university life. They criticized those men, supporters of women's aspirations, who would surrender "our universities to the invasion of women ... thereby falsifying their entire character." Universities were more than just centers of learning. They fostered friendships and built character in young men. Women would introduce a discordant note.

In 1898 the Academic Senate of the University of Erlangen went officially on record that the admission of women students was a "measure that would overthrow all academic order." Slowly, however, the barriers began to come down. In

1901 the Universities of Freiburg and Heidelberg in the State of Baden admitted women on an equal footing with men in all departments. Other German states followed Baden's example, and by 1908 co-education became generally accepted at German universities.

One major difficulty remained, however, for any woman who wished to qualify for enrollment at a university. In general, women could not attend a *gymnasium*, or secondary school, to get the rigorous preparation required for university admission. Instead they could take the state matriculation examination to demonstrate their proficiency for their chosen course of study. Emmy Noether, having decided to continue her studies at the university, began to study for this examination. She was able to get permission to audit courses at the University of Erlangen from 1900 to 1902, and in July 1903, she took and passed the matriculation examination which was given at the *Realgymnasium* in Nürnberg.

During the winter semester 1903–1904, she attended the University of Göttingen, whose mathematics faculty were known to be more liberal in their policies towards women students than other German institutions. But in the fall of 1904, Emmy returned to Erlangen and was enrolled as the only woman in the faculty of mathematics with 46 male students. She worked under Paul Gordan and on December 13, 1907, successfully defended her thesis, receiving a summa cum laude. She was awarded her degree in July 1908.

Emmy Noether belonged to the small group of pioneering women who were at that time rejecting the traditional decorous role, assigned to their peers in the middle and upper classes, to knock at the doors of academia. When she attended the University of Erlangen there were only three other women regularly enrolled: a Russian studying philosophy, and two other German girls, one in medicine, and the other in languages. What inspired Emmy Noether in 1900 to diverge from the norm? Why did she decide that she did not want either to stay home or to teach French and English to other well-brought-up young ladies? Was it the cumulative effect of years of having been exposed to mathematical talk by the Erlangen mathematicians who frequented the Noether home? Was it the example of her younger brother Fritz, now beginning his university studies, and undoubtedly full of enthusiasm for mathematics which he probably discussed at home with his father? Was it a sudden rebellion against the prescribed course, a desire to be herself, to follow her own inclination? Herman Weyl, a close friend and co-worker at the University of Göttingen, stressed in his commemorative address after her death in 1935, that Emmy had never been a rebel in her life, but who knows her inner thoughts in the early 1900s? We shall probably never know and can only speculate. What matters is that she did take the step, she did persist, despite all the odds against women, and did go on to become one of the most distinguished algebraists of her century.

Her university training completed, Emmy Noether worked without a formal appointment or any monetary compensation at the Mathematical Institute in Erlangen, partly helping her father and partly on her own research. Slowly recognition began to come her way. In 1908 she was elected to membership in the *Circolo Matematico di Palermo* and in the following year she was invited to join the *Deutsche Mathematiker Vereinigung* and began to give lectures at its meetings.

She first addressed the Salzburg meeting of the Society in 1909, and in 1913 read a paper in Vienna. By 1915 she had published some half-dozen papers which demonstrated an impressive knowledge of certain aspects of mathematics. In that year recognition of her work came from Felix Klein and David Hilbert, who felt that her research complemented their own work on relativity theory and invited her to join the Mathematical Institute in Göttingen. She accepted, moved to Göttingen, and remained there until forced out by the Nazis in 1933.

Having called Emmy Noether to Göttingen, Hilbert and other mathematicians wanted a regular faculty appointment for her, or, at the very least, that she be allowed to get the *Habitilation* and thus become a *Privatdozent*. But permission for the *Habitilation* had to be obtained by vote of the entire philosophy faculty, in which were included not only scientists and mathematicians, but also philosophers, philologists, and historians, most of whom were unswervingly opposed to permitting a woman to try for the *Habitilation*. The discussion regarding Emmy Noether at the faculty meeting has become a classic story. Opponents asked: "How can we permit a woman to become a *Privatdozent*? Having become a *Privatdozent*, she can then become a professor and a member of the University Senate. Is it possible that a woman enter the Senate?" Further they asked, "What will our soldiers think when they return to the University and find that they are expected to learn at the feet of a woman?"

Hilbert answered bluntly: "I do not see that the sex of the candidate is an argument against her admission as a *Privatdozent*. After all, the University Senate is not a bathhouse." Despite all the arguments advanced by the mathematicians, the faculty rebuffed the nomination. Hilbert finally solved the issue by having Emmy Noether lecture in his stead. In fact, the catalog for the winter semester of 1916–1917 at Göttingen contains the following entry: "Mathematical physics seminar: Professor Hilbert, with the assistance of Dr. E. Noether, Mondays, from 4–6, no tuition."

So passed the war years. Finally, November 1918 saw the end of both the war and the monarchy in Germany. The changes in the outward political structure of Germany did not alter the conservative bent of the men in academia, but they forced some concessions. Emmy Noether was finally allowed to take her *Habitilation* in June 1919. It should be pointed out, however, that the title *Privatdozent*, to which the *Habitilation* entitled her, did not carry a regular stipend. Emmy Noether —by now a woman in her late thirties, with a growing scholarly reputation— still had no secure position in the academic world. Eventually, she was given a *Lehrauftrag*, or lectureship, in algebra to which was attached a small stipend. She could now teach officially, give examinations, and direct dissertations. But she was not, as we would say in the United States, on the tenure track. Unlike German university professors with regular appointments, she was not a member of the state civil service, and enjoyed none of the fringe benefits or pension rights associated with regular positions.

Even though her mathematical work became widely known, her status at the University of Göttingen never changed. Mathematical recognition came at the International Mathematical Congresses in 1928 and 1932. In 1928, at Bologna, she was the main speaker at one of the Congress' section meetings, and in 1932, at

Zurich, she addressed the plenary session of the Congress. In that same year, together with Emil Artin, she received the Alfred Ackermann-Teubner Memorial Prize for the Advancement of Mathematical Knowledge.

The rise of Nazism in Germany shattered the life and work of many German scholars, among them Emmy Noether. In April 1933, the purge of Jews and persons of Jewish descent forced the retirement of all civil servants with at least one Jewish grandparent.

Emmy, deprived of her modest stipend, unable to participate officially in the academic life of the University at Göttingen, accepted the offer of a guest professorship at Bryn Mawr.

After the Second World War, German academia tried to make amends for its shabby treatment, and a new generation of German scholars recognized Emmy Noether's contributions to mathematics. In 1958 the University of Erlangen brought together many of her former students and their students to memorialize the fiftieth anniversary of her degree and to discuss her work, its applications and its influence on subsequent research. In 1960 the city of Erlangen named one of the streets in a new residential neighborhood the Noetherstrasse. On February 27, 1982, at a Festkolloquium a Noether memorial tablet was unveiled in the Mathematical Institute of the University of Erlangen.

And next week, I shall be going to Erlangen to participate in the dedication of the new Emmy-Noether-Gymnasium, a school for both boys and girls, which will emphasize mathematics and the natural sciences along with modern languages, subjects which at one time or other have played an important role in Emmy's life. I think that Tante Emmy would have approved.

Emmy Noether in Bryn Mawr

GRACE S. QUINN,* RUTH S. MCKEE,† MARGUERITE LEHR‡ and OLGA TAUSSKY§

Do you wonder how Miss Noether herself would react to this celebration in her honor? We can imagine her turning or tilting her head to one side, considering the import of it all and smiling shyly, but proudly, her eyes bright behind those thick lenses. She would be listening intently to these algebra talks, never missing a word, perhaps a little breathless with concentrated interest.

After four years of college teaching and a published doctoral thesis under C. C. MacDuffee, I came to Bryn Mawr on a fellowship in the fall of 1934, the beginning of Miss Noether's second year in the United States. This grant was called the Emmy Noether Fellowship and it paid all of my expenses: room, board, and tuition.

There were four of us students: Marie Weiss had come from teaching at Newcomb College, Tulane University, having studied with Manning at Stanford, to accept the Emmy Noether Scholarship; Olga Taussky had postponed a grant from Girton College, Cambridge, and had come from a research post in Göttingen to take advantage of the opportunity to study with Miss Noether on the Bryn Mawr Foreign Scholarship; Ruth Stauffer was continuing her studies on a Bryn Mawr scholarship leading to the doctorate; and myself. We not only studied together, attended Miss Noether's and Mrs. Wheeler's lectures also, but we really played together, walking down Gulph Road with Miss Noether in the lead discussing mathematics intensely all the while unmindful of the traffic or birding through the woods with Mrs. Wheeler, binoculars in hand, attending teas hosted by department chair Mrs. Wheeler in her comfortable Low Buildings faculty apartment or even by one of us in her own room in the graduate dormitory, Radnor Hall.

* 4801 Jamestown Road, Bethesda, MD 20816, U.S.A.
† 201 Gurney, Pennswood Village, Newtown, PA 18940, U.S.A.
‡ Conwyn Arms, Apt. 309, Bryn Mawr, PA 19010, U.S.A.
§ Department of Mathematics, 253-37, CALTECH, Pasadena, CA 91125, U.S.A.

Two other graduate students in mathematics, Madeline Levin and Frances Rosenfeld, also participated occasionally.

With only a few exceptions our attention in Miss Noether's class was devoted to working through the book *Class Field Theory* by Helmut Hasse. We also had our own interests. I began working on material suggested by Miss Noether which resulted in a paper "Number theory in a hypercomplex system" presented to the Society in April 1936. Late in the winter of 1934–1935 I was pleased to receive an appointment to teach in a nearby school for the following year thus enabling me to continue my studies with my renowned professor. Unfortunately, my plans were shattered by her untimely death.

Forty seven years have elapsed! I have been accused of having the memory of an elephant. I am not so sure. However, there are some events which leave a deep impression.

When asked to participate on this panel I sought advice and received some suggestions in the form of queries which I'll try to answer.

1. Did you have much contact with her? Yes, indeed, as already implied.
2. What was her influence on undergraduate students? There was no opportunity to observe. As far as I know neither Miss Noether nor we four students even knew[1] any undergraduate mathematics students. It is possible that Mrs. Wheeler was holding such plans for the future.
3. Did you get a strong feeling about her great love for mathematics? No doubt. Yes! Yes!
4. What was her influence on the graduate students? That was really pervasive as far as we four were concerned but I can not answer for the other two.
5. On the faculty in the Mathematics Department. Another member of the panel can answer that question better than I.
6. Did you ever go with her to Princeton and, if so, what were your experiences there? Yes, we did, and to the October meeting of the Society in New York and to the biweekly graduate seminar or club at the University of Pennsylvania Department of Mathematics.

(We are indebted to Professor Alice T. Schafer of Wellesly College for having formulated the above queries.)

I do recall that she missed the annual meeting of the Society in Pittsburgh that year which Marie, Olga and I did attend. We usually traveled to Philadelphia and to Princeton in Mrs. Wheeler's car, although Miss Noether often went to Princeton by rail for her lectures. I recall that when Richard Brauer and his wife came to visit, staying at Mrs. Wheeler's, somehow or other I was honored by being invited to accompany Miss Noether and the Brauers, all to go with Mrs. Wheeler in her car, to visit the Schoenbergs in Swarthmore.

Much seems to blend together in one big blur but some events standing out clearly are the following: our trip together, Miss Noether and about three of us, to attend the fall Society meeting at Columbia University, on the trains and

[1] Since writing the above I have been reminded that Mrs. Wheeler referred an undergraduate to Olga for special instruction on a project.

especially on the New York subway where her enthusiasm was unselfconsciously exhibited; a visit early in the year to her rooms in a house south of the Lancaster Pike to view her furniture which she had shipped from Germany and of which she was really proud, especially of her massive desk.

The events of the last 15 days of her life, March 31 to April 14, made a lasting impression. At that time the college had the usual spring break. The dorm closed. We dispersed. I stayed with friends nearby. Olga took a room at Atlantic City. Miss Noether invited me to accompany her to visit Olga and Atlantic City for the day, Sunday, March 31. We three had an interesting time. A long-time friend who had come from Germany to live here came to visit me over the next weekend. I took her in the early evening on Sunday, April 7, to meet Miss Noether. The two German women enjoyed a pleasant conversation in their native tongue about their homeland.

On the next day, Monday, April 8, Mrs. Wheeler summoned us. We were informed that Miss Noether would enter the Bryn Mawr Hospital that day for the removal of a uterine tumor. The next day, the day before her operation on Wednesday, we four called on her in her hospital room. We had a simple gift, a copy of the magazine *Town and Country*, I believe it was called. Later we learned that her operation had been successful and that she was proud that no more had to be removed than the alien tumor itself. On Saturday when we tried to call on her we learned that she was not yet able to receive visitors. On Sunday when we were all in the dormitory Ruth was called to the phone. Mrs. Wheeler informed her that Miss Noether had just expired rather unexpectedly from an embolism, just when we were anticipating her recovery.

This was the first time that some of us had ever experienced the loss of someone close. The effect was traumatic: shock, sleeplessness.

In the middle of the next week there was a Quaker memorial service in the living room of the president's residence. We heard music played softly by a string ensemble in a nearby room. A closed black box along the side of the room reminded us of the loss of our beloved professor. Four eulogies were scheduled: Mrs. Wheeler represented Miss Noether's American colleagues; Richard Brauer, speaking in German, her German colleagues; Ruth, her American students; and Olga, her foreign students.

We remember her for her brilliant mind, her beautiful character, her indomitable courage, her democratic approach, her love of life and her love of mathematics, in particular.

In the summer of 1936 I attended the International Congress of Mathematicians in Oslo where I met Miss Noether's brother Fritz who had traveled from Tomsk, Siberia, to give a paper. On the way to Oslo I stopped in Göttingen. Because Richard Courant had kindly furnished me with letters of introduction I was invited to attend a few lectures of Helmut Hasse in his class room and to call on the David Hilberts in their home. On my second visit there Frau Hilbert showed me a collection of photos of her husband's students. It was interesting to note the pictures of Americans who had studied with Hilbert, especially that of one of my former professors. These encounters are mentioned because, of course, these folks were eager to learn about Miss Noether's life in the United States.

Some of my own pictures and other mementos have been placed in an album. Because this is an association of women I have added pictures of other women mathematicians and a few males. The album may be examined by those interested. The necklace around my neck is from the estate of Emmy Noether.

At this juncture I shall remove the necklace from my own neck and present it to her nephew, Dr. Gottfied Noether, for his only daughter Monica, for her to own to remember her illustrious great-aunt Emmy.

GRACE S. QUINN

I would like to share with you my memories as a student of Emmy Noether. Not only did she teach us abstract algebra but we learned from her how to apply her methods to our work and our living. First let me tell you how it was with us at Bryn Mawr that first year. As I remember there were five of us who had never been exposed to any abstract algebra. She started by giving us a short assignment in the first volume of Van der Waerden's *Moderne Algebren.* A day or two later she stopped by the seminar room and asked me how it was going. "Well," said I, "I'm having trouble knowing how to translate all these technical terms such as Durchschnitt." "Ah-ha," she said, "don't bother to translate, just read the German." That is the way our strange method of communication began. Although we students were far from conversant with the German language, it was very easy for us to simply accept the German technical terms and to think about the concepts behind the terminology. Thus from the beginning we discussed our ideas and our difficulties in a strange language composed of some German and some English.

Miss Noether's classes were not lectures, they were discussions. Proofs were sometimes presented by us and sometimes suggested by Miss Noether. The strange phenomena, as I look back on it, was that from our point of view, she was one of us, almost as if she too were thinking about the theorems for the first time. There was lots of interest and competition and Miss Noether urged us on, challenging us to get our nails dirty, to really dig into the underlying relationships, to consider the problems from all possible angles. It was this way of shifting perspectives that finally hit home. I must admit that Miss Noether was not the first of my professors who had tried but suddenly the light dawned and Miss Noether's methods were the only way to attack modern algebra. Miss Noether was a great teacher!

Miss Noether's methods of working and thinking became the basis for my analytical work for the research agency of the Pennsylvania State Legislature for almost 30 years. It is probably heresy for me to mention this in front of so many theoretical mathematicians but there is a great need in government for abstract, imaginative thinkers to help solve all sorts of problems. For example: What are the basic cost factors in a given government funded program? What is the taxpayer's money really accomplishing? During my career we searched for answers to these questions in such areas as the construction of public school buildings, the operation of State mental hospitals, the faculty workloads at various levels of education, highway engineering as directed toward traffic safety. We chewed over the characteristics and searched for the basic independent variables when considered from all possible points of view. Other times the problem was to find the relevant variables to determine an equitable distribution of appropriations. What was the most important factor? population density? financial need? or, simply, geography?

These good old-fashioned "reading problems" were, from my point of view, a perfect area to apply Miss Noether's methods of working and thinking, keeping an open mind to the possible angles and at the same time understanding the basic factors of the problem. Although I know the schools are desperate for good mathematics teachers, there may be some who simply do not want to teach. To these, I would like to make a plea, call it a commercial, if you like; the public sector is also desperate for abstract imaginative thinkers who always did like "reading problems."

As I remember Miss Noether's methods of thinking and working, they were simply a reflection of her way of life; that is, recognize the unessentials, brush them aside and enter whole heartedly into the present. This was, as you all know, far from a superficial achievement on the part of Miss Noether. Bitterness and jealousy were rejected by her as unessential. There was never any indication of bitterness toward Germany even though Hitler's government denied her the right to teach at Göttingen. Nor was there any sign of jealousy because of her treatment as a woman even in the end when her colleagues from Göttingen went to the Institute for Advanced Study at Princeton with possibilities of many promising advanced young students. Her lot, instead, was Bryn Mawr, with a Mathematics Department consisting of four faculty members headed by Mrs. Wheeler and five graduate students who had never been exposed to any abstract algebra. She knew, of course, that there was a possibility of trips to Princeton for lectures and discussions.

It continues to amaze me that there appeared to be no problems confronting her life at Bryn Mawr. Her greatest pleasure seemed to us to be happily discussing various mathematical ideas. I remember one time she stopped by the seminar room and she and I started talking about an idea that had struck me. We became so involved we talked right through the dinner hour, and dinner meant a lot to both of us!

From our point of view she was a happy, well-adjusted person, very much interested in getting to know Mrs. Wheeler, her faculty, and the students. She was eager to fit into the social life at Bryn Mawr. She knew, for example, that Mrs. Wheeler entertained the students and faculty in her apartment for tea so she wanted to have a tea party in her home. I must tell you that her living arrangements were modest but comfortable. She had a room and boarding arrangement not far from the campus with Mrs. Hicks who was a kind, thoughtful person who took a great interest in taking care of Miss Noether. Well, Mrs. Hicks planned a lovely tea party and Miss Noether asked Mrs. Wheeler to preside at the tea table. The setting was complete, the guests arrived, and Miss Noether beamed happily; but soon she was noticeably upset and went out to the kitchen for Mrs. Hicks. It was obvious to Miss Noether that pouring tea, rather than being an honor, was an onerous job; and she had asked Mrs. Hicks to pour so that her good friend could enjoy herself at the party. Once again all was sunshine and light. In other words, correct an apparent problem in the simplest way.

Another example of Miss Noether's way of life was demonstrated one afternoon during her second year at Bryn Mawr when she was walking with her four students. She liked walking in the country so we started across an open field behind the college. I soon realized that we were heading straight for a rail fence. Miss Noether was immersed in a mathematical discussion and went merrily along, all of us walking at a good clip. We got closer and closer to the fence. I was apparently the weak sister, concerned mostly in how we would handle the fence. For those of us in our twenties it would be no problem but, from my point of view, however would this "old lady," fiftyish, handle the fence? On we marched right up to the fence and without missing a word in her argument she climbed between the rails and on we went.

Those of us who knew Miss Noether will be forever indebted to her for her example of selfless living. As Hermann Weyl said:

Her heart knew no malice
She did not believe in evil.

RUTH S. MCKEE

In the afternoon of April 17, 1935, after Miss Noether's death in the Bryn Mawr Hospital, President Park asked a little group of mathematicians to meet at her house. They were all Göttingen connected in some way: Richard Brauer, Olga Taussky, Anna Pell Wheeler. President Park asked me, not at all an algebraist or Göttingen, to speak next morning at chapel for five minutes, on the impact which Miss Noether's two years here had made on a young American, as I was. Not until a month ago did I know that the little handwritten three pages of that talk were in her files, and then transferred, by a careful archivist, to my file. I asked at once if I might be the last speaker on this panel, and, instead of reminiscing, read what I wrote in 1935.

"At the opening Convocation in 1933, President Park announced the coming of a most distinguished foreign visitor to the Faculty, Dr. Emmy Noether. Among mathematicians that name always brings a stir of recognition; the group in this vicinity waited with excitement and many plans for Dr. Noether's arrival. At Bryn Mawr there was much discussion and rearrangement of schedule, so that graduate students might be free to read and consult with Miss Noether until she was ready to offer definitely scheduled courses. For many reasons it seemed that a slow beginning might have to be made; the graduate students were not trained in Miss Noether's special field—the language might prove a barrier—after the academic upheaval in Göttingen the matter of settling into a new and puzzling environment might have to be taken into account. When she came, all of these barriers were suddenly non-existent, swept away by the amazing vitality of the woman whose fame as the inspiration of countless young workers had reached America long before she did. In a few weeks the class of four graduates was finding that Miss Noether could and would use every minute of time and all the depth of attention that they were willing to give. In this second year her work had become an integral part of the department; she had taken on an honors, i.e., undergraduate senior student, her group of graduates has included three research fellows here on scholarships or fellowships specially awarded to take full advantage of her presence, and the first Ph.D. dissertation directed at Bryn Mawr by Miss Noether has just gone to the Committee bearing her recommendation.

"Professor Brauer in speaking yesterday of Miss Noether's powerful influence professionally and personally among the young scholars who surrounded her in Göttingen said that they were called the Noether family, and that when she had to leave Göttingen, she dreamed of building again somewhere what was destroyed there. We realize now with pride and thankfulness that we saw the beginning of a new "Noether family" here. To Miss Noether her work was as inevitable and natural as breathing, a background for living taken for granted; but that work was only the core of her relation to students. She lived with them and for them in a perfectly unselfconscious way. She looked on the world with direct friendliness and unfeigned interest, and she wanted them to do the same. She loved to walk, and many a Saturday with five or six students she tramped the roads with a fine disregard for bad weather. Mathematical meetings at the University of Pennsylvania, at Princeton, at New York, began to watch for the little group, slowly growing, which always brought something of the freshness and buoyance of its leader.

"Outside of the academic circle, Miss Noether continually delighted her American friends by the avidity with which she gathered information about her American environment. She was proud of the fact that she spoke English from the very first; she wanted to know how things were done in America, whether it were giving a tea or taking a Ph.D., and she attacked each subject with the disarming candor and vigorous attention which won everyone who knew her.

"Emmy Noether might have come to America as a bitter person, or a despondent person. She came instead in open friendliness, pleased beyond measure to go on working as she had, even in circumstances so different from the ones she had loved. And our final consolation is that she made here too a place that was hers alone. We feel not only greatly honored that she wanted to stay and work with us; we feel profoundly thankful for the assurance that her friends have brought us—that her two short years at Bryn Mawr were happy years."

Chapel MARGUERITE LEHR
Thursday, April 18, 1935

Meeting Emmy Noether was one of the great things in my life. It was not just meeting her. Twice I spent almost a whole year at the same place. She was a teacher and she had a great urge to make people see her methods and to understand them. So I picked up a great deal from her and from the other people who worked more closely with her. In Göttingen her favorite student Deuring spoke frequently to me, he also gave me copies of his publications. At the time I was there he was more of a colleague to her than a student and was in fact allowed to say "Du" to her, a great honour in Germany. Then there was Witt, not yet her student, but on his way to become her student. Then there was Schilling, her Ph.D. student, with a thesis whose title is the same as the title of my lecture in this symposium. Frequently her older colleagues like Hasse, van der Waerden visited her and I had contact with them. A former girl student of hers was frequently mentioned, Hermann. Her work on effective methods is very much appreciated nowadays. Saunders MacLane was greatly influenced by Emmy towards his later work in cohomology.

But it was not only the people who were in Göttingen during my stay from whom I learnt about her work. At that time she had developed her whole power and influence over many subjects. That particular year was perhaps the happiest in her life. Her influence ranged over many subjects and one of them is topology. Here Alexandroff, her greatest admirer, expressed his indebtedness to her algebraic methods in print and later Lefschetz said to me personally that he wished he had known these methods earlier.

At Bryn Mawr it was particularly easy for me to profit also from her school. There was her thesis student Ruth. There was Marie Weiss who worked on a problem explicitly suggested to her, namely units in cyclic fields, using ideas of Latimer. For this we had to thank Grace. For she was a student of MacDuffee who worked closely with Latimer. Grace introduced us to her teacher and Emmy thought highly of his work. His work and also the work of Grace are mentioned in Deuring's book on algebras. But there is one paper he missed there, it is written by both Latimer and MacDuffee. They both sent me a reprint, but somebody went off with one copy. I wrote many papers using this and so now do others.

Olga Taussky

The Study of Linear Associative Algebras in the United States, 1870–1927

JEANNE LADUKE*

When the "modern algebra" of Emmy Noether and her school became known in the United States in the early 1930s it encountered and interacted with algebra as it had been known and practiced here. It should be useful to examine one aspect of that algebra in order to prepare a background for understanding the relationship of Noether's work to that of algebraists in the United States. One area in particular, the study of hypercomplex number systems, or linear algebras as they came to be called, had a particularly American flavor in its development from 1870 to about 1927. For this reason and because much of Noether's most influential work is related in some way to hypercomplex number systems, I shall examine this aspect of algebra. I will describe how hypercomplex numbers were viewed and what language was used to describe them, what were seen as major problems, and what principles guided American algebraists in this field during the period 1870–1927.

For now we consider a hypercomplex number system to be a finite-dimensional vector space over the reals with a multiplication which is associative and which distributes over addition. Hypercomplex numbers were the result of extending the concept of number from that of real number, to complex number, to quaternion, to "hypercomplex number." The study of hypercomplex numbers was initiated by the discovery of quaternions in 1843 by William Rowan Hamilton (1805–1865) in his search for a 3-dimensional analogue of complex numbers. Unable to find a 3-dimensional system which acted like ordinary complex numbers, that is, in which division by non-zero values was always defined, he did find a 4-dimensional one which satisfied this division requirement, except that commutativity of multiplication failed. The theory of quaternions, their interpretations, and applications were developed to a high degree, especially in Great Britain. Hamilton devoted the rest of his life to their study; his *Lectures on Quaternions* appeared in 1853 and his two-volume *Elements of Quaternions* in 1866, the year after his death. Peter

* Department of Mathematical Sciences, DePaul University, Chicago, IL 60614, U.S.A.

Guthrie Tait (1831–1901), at the University of Edinburgh from 1860 until his death, was especially interested in promoting the application of quaternions to physical problems. His *Elementary Treatise on Quaternions* was published in three editions beginning in 1867. In the preface, dated 1873, to the second edition of this work he remarked that he had "many pleasing proofs that the work has had considerable circulation in America."

Other systems of hypercomplex numbers were sought and found. Hamilton's quaternions were always considered to have real coefficients, but in his *Lectures* he also considered them with complex coefficients, and called these biquaternions. By early 1844 John T. Graves had written to Hamilton about his theory of "octaves" or "octonions," an 8-dimensional system in which associativity as well as commutativity of multiplication failed. They are now known as the Cayley numbers since a description of them first appeared in print in a postscript to a paper on elliptic functions by Cayley in 1845.

The study of quaternions and their generalizations was taken up almost immediately by Benjamin Peirce (1809–1880) in the United States. As early as 1848 Peirce included quaternions in his lectures at Harvard. In 1870 he read a memoir before the National Academy of Sciences in Washington, D.C. entitled "Linear associative algebra." Originally only a few copies were published in lithograph, but the address was published in the *American Journal of Mathematics* "with Notes and Addenda, by C. S. Peirce, Son of the Author" in 1881, the year after Benjamin Peirce's death. A note preceding the article indicated:

> This publication will, it is believed, supply a want which has been long and widely felt, and bring within reach of the general mathematical public a work which may almost be entitled to take rank as the *Principia* of the philosophical study of the laws of algebraical operation.

Peirce noted early in the paper that "the language of algebra has its alphabet, vocabulary, and grammar." The alphabet consists of letters such as i, j, k which he called fundamental conceptions and by which he meant what we call elements of a basis. The vocabulary consists of various signs such as $+$, $-$, $\times$; and the grammar "gives the rules of composition by which the letters and signs are united into a complete and consistent system." He described explicitly the distributive principle and the associative and commutative principles for multiplication. Peirce then introduced the nomenclature "linear algebra" and defined a linear algebra as "an algebra in which every expression is reducible to the form of an algebraic sum of terms, each of which consists of a single letter with a quantitative coefficient." He studied linear associative algebras with complex coefficients.

Peirce then proceeded to give a detailed analysis of the structure of such algebras. He introduced the concepts "nilpotent" and "idempotent" and showed that in any linear associative algebra either there is an idempotent or every expression in the system is nilpotent. Then he demonstrated that in any algebra containing an idempotent, the units (basis elements) may be chosen so that they fall into four disjoint sets defined with respect to the idempotent. This yields what is now called the "Peirce decomposition" of an algebra. Peirce's analysis is followed by a 97-page "investigation of special algebras" in which his goal was to enumerate essentially all complex associative linear algebras of dimension 6 or less. These algebras are described by displaying the multiplication tables of the basis elements. We shall examine the reception and influence of Peirce's work on linear algebras later in this paper.

Two notes by C. S. Peirce which appeared as addenda to the 1881 paper are also notable. One describes the "relative form" of an algebra and shows that "any associative algebra can be put into relative form, i.e. that every such algebra may be represented by a matrix." In the second note he proved, apparently independently of Frobenius, and three years later, "that ordinary real algebra, ordinary algebra with imaginaries, and real quaternions are the only associative algebras in which division by finites always yields an unambiguous quotient."

In the latter part of the nineteenth century much of the work on hypercomplex number systems in the United States was done by the Englishman James Joseph Sylvester (1814–1897) who between 1877 and 1883 was at Johns Hopkins. Sylvester, in a series of papers, particularly from 1882 to 1884, examined the connections between hypercomplex numbers and the theory of matrices which had been developed earlier by Cayley. Some of Sylvester's work appeared in the *Comptes Rendus* and thus came to the attention of continental mathematicians, some of whose work we will sketch briefly later.

From the turn of the century until about 1907 there were many different threads of activity in the United States in the study of linear associative algebras. There was a reassessment of Peirce's methods; there was an interest in postulational systems; there were the beginnings of the Wedderburn structure theory; and finally there were efforts to describe and characterize various kinds of algebras. Meanwhile Eduard Study (1862–1930) and Georg Scheffers (1866–1945) working on the continent in the late nineteenth century and using methods entirely different from Peirce's, namely, methods based on the theory of continuous groups of Sophus Lie (1842–1899), studied hypercomplex numbers. In particular, they enumerated hypercomplex number systems in a small number of units.

In 1902, H. E. Hawkes (1872–1943) of Yale published two papers in which he analyzed Peirce's methods and made more rigorous Peirce's work on enumeration and classification of linear associative algebras of low dimension. His "Estimate of Peirce's linear associative algebra" begins:

> . . . [Peirce] made the first systematic attempt to classify and enumerate hyper-complex number systems. Though his work attracted wide and favorable comment in England and America at the time, continental investigators on the subject during the last fifteen years have given him scarcely the credit which his results and his methods deserve. Adverse criticism has been due in part to a misunderstanding of Peirce's definitions, in part to the fact that certain of Peirce's principles of classification are entirely arbitrary and quite distinct in statement from those used by Study and Scheffers, in part to Peirce's vague and in some cases unsatisfactory proofs, and finally to the extreme generality of the point of view from which his memoir sprang, namely a 'philosophic study of the laws of algebraic operation.'

In that paper Hawkes described principles of classification used by Peirce either explicitly or implicitly and compared them with those employed by Study and Scheffers. He noted that the theorems stated by Peirce were all true although in some instances the proofs were invalid. In a second paper, "On hypercomplex number systems" published in the same year, Hawkes used Peirce's methods and the theory of transformation groups to prove again Peirce's main results and to enumerate all number systems of the type Scheffers had considered.

Henry Taber (1860–1936) re-examined Peirce's theorems in 1904. He noted that whereas Hawkes had taken advantage of the theory of transformation groups to establish one of Peirce's theorems, he wanted "to establish Peirce's method without recourse to the theory of groups." Thus, the interest in Peirce's work and the problem of enumerating linear associative algebras of low dimension received much attention well into the twentieth century.

While Hawkes at Yale and Taber at Clark were re-establishing the work of Peirce, the University of Chicago, soon after its beginning in 1892, supplanted Johns Hopkins as the leading American research center in mathematics. Under the direction of E. H. Moore (1862–1932) the department at Chicago exerted a major influence on the American mathematical community. In particular, Moore emphasized both axiomatic systems and generalization. Moore's first student, Leonard Eugene Dickson (1874–1954), who joined the faculty at Chicago in 1900, was to become in many respects the most influential algebraist in the country prior to about 1925. Dickson (1903b) gave definitions of a linear associative algebra by independent postulates in which he generalized the field of scalars to be any abstract field. In 1905 he pursued a further generalization (1905b). These papers were part of a general interest at the beginning of the century in postulate systems. For example, at roughly the same time papers appeared in the *Transactions* on definition of a field by independent postulates (Dickson 1903a, 1904; Huntington 1903), on the projective axioms of geometry (Moore 1902a), and on a definition of abstract group (Moore 1902b).

The University of Chicago, center of algebraic activity in the United States at the turn of the century, also played a significant role in the development of the Wedderburn structure theory,[1] for it was to Chicago that Joseph Henry Maclagan Wedderburn (1882–1948) of Scotland came as a Carnegie Fellow for the year 1904–05. In the late nineteenth century work on the structure of hypercomplex systems over the complex numbers was done by Molien and Cartan in Europe. The paper "Hypercomplex numbers, Lie groups, and the creation of group representation theory" by Thomas Hawkins (1972) includes a detailed description of this theory. Hawkins wrote (p. 256):

> Theodor Molien (1861–1941) and Elie Cartan (1869–1951) made essentially the same discoveries about the structure theory of hypercomplex systems—discoveries that went far beyond the work of Study and Sheffers. Of special importance from our perspective are the introduction of the notions of simple and semisimple systems, the characterization of the former as complete matrix algebras, and the discovery of necessary and sufficient conditions for semisimplicity in terms of the nonsingularity of certain bilinear or quadratic forms. . . . their methods were entirely different, and it appears that they were led to the same results independently of each other.

[1] Karen Parshall gives an extensive description and analysis of the structure theory of Wedderburn and its antecedents in a recent Ph.D. dissertation (1982).

In 1905 there appeared a joint paper, "On the structure of hypercomplex number systems," by Saul Epsteen, then at the University of Chicago, and by J. H. Maclagan-Wedderburn. They observed:

> In this paper theorems regarding linear associative algebras are enunciated which are analogous to the Jordan–Hölder theorems concerning the quotient groups and indices of composition in the theory of finite groups . . . and of the corresponding Vessiot–Engel theorems in the theory of continuous groups. . . . The methods used throughout are rational and hence the results apply to hypercomplex number systems in which the coefficients are restricted to be marks of a given field, finite or infinite.

They cited work by Frobenius and Cartan on the theory of "complexes" and invariant complexes (two-sided ideals) and work by Molien on the notion of an "accompanying system." Their use of the theory of complexes and the fact that the coefficients were allowed to be in an arbitrary field were both characteristic features of the structure theory to be presented by Wedderburn in 1907.

A fourth line of inquiry in the period prior to 1907 was the effort to describe and characterize certain types of algebras, especially division algebras. Wedderburn and Dickson were key figures in this effort. Both were interested in finite algebras and were engaged with the question at Chicago early in 1905. Wedderburn's result that any finite linear associative algebra which is also a division algebra (an algebra in which division by non-zero elements is uniquely possible) is a field (and, therefore, is a Galois field) appeared in the *Transactions* in 1905. Wedderburn gave three proofs of this major theorem. A long paper by Dickson, "On finite algebras," appeared in the *Göttingen Nachrichten* later that same year (1905a). It includes the above theorem on finite algebras and cites the Wedderburn paper, noting in a footnote, however, that two of Wedderburn's proofs were based on a prior theorem of Dickson. The major part of Dickson's 1905 paper is devoted to examining the independence of a set of postulates for a finite field and to determining two types of finite algebras which fail to satisfy all the field postulates: first, division algebras in which commutativity of multiplication and the right-hand distributive law fail; second, non-associative division algebras in which multiplication is commutative. Further work by Dickson on division algebras over arbitrary fields followed immediately (1906a, 1906b).

The year 1907 saw both an effort to sum up the past achievements in linear algebras and what would much later be seen as a ground-breaking work in new directions. First was the appearance of the volume, *Synopsis of Linear Associative Algebra*, by James Byrnie Shaw (1866–1948), at the time a professor at James Millikin University. The volume consists mainly of a collection of definitions and theorems (mostly without proofs) with a large section devoted to the enumeration of various algebras. In his introduction, Shaw summarized two views of linear algebra. One, he said, sees the elements of an algebra as a pair, triple, etc., "seeks to derive all properties from a multiplication table," and is the result of an "attempt to base all mathematics on arithmetic." The second point of view which he advocated "regards the number in a linear algebra as a single entity," seeks to derive the properties "from definitions applying to all numbers of an algebra," and is the result of "the attempt to base all mathematics on algebra, or the theory of entities defined by relational identities." Still referring to the second view he continued:

> Such definition of algebra, or of an algebra, is a development in terms of what may be called the fundamental invariant forms of the algebra. The characteristic equation of the algebra and its derived equations are of this character . . . .

Shaw's views of algebra and methods of investigation, while useful for describing the past and, indeed, some of the work in linear algebras in the immediate future, did not foresee the directions to be taken by Wedderburn and later by Emmy Noether and her school.

I indicated above that the seeds of the Wedderburn structure theory appeared in a paper by Epsteen and Wedderburn in 1905. This structure theory came to fruition in Wedderburn's paper "On hypercomplex numbers" (parts of which were read in the Mathematical Seminar of the University of Chicago early in 1905) which is dated November 14, 1907 in the *Proceedings of the London Mathematical Society*. Wedderburn began the paper:

> The object of this paper is in the first place to set the theory of hypercomplex numbers on a rational basis. The methods usually employed in treating the parts of the subject here taken up are, as a rule, dependent on the theory of the characteristic equation, and are for this reason often valid only for a particular field or class of fields. Such, for instance, are the methods used by Cartan . . . .
>
> My object throughout has been to develop a treatment analogous to that which has been so successful in the theory of finite groups. An instrument towards this lay to hand in the calculus developed by Frobenius . . . .
>
> Most of the results contained in the present paper have already been given, chiefly by Cartan and Frobenius, for algebras whose coefficients lie in the field of rational numbers . . . .

The algebras (called "algebras" rather than "linear associative algebras" at the suggestion of Dickson) under consideration by Wedderburn are finite-dimensional algebras over *any* field. Wedderburn summarized his principal theorem: "Any algebra can be expressed as the sum of a nilpotent algebra and a semi-simple algebra." (p. 109). His other major results were: any semi-simple algebra, which is not simple, is the direct sum of simple algebras; and furthermore, any simple algebra can be expressed as the direct product of a primitive algebra (he means division algebra) and a simple matrix algebra. Here an algebra is simple if it contains no invariant subalgebra, semi-simple if it contains no nilpotent invariant subalgebra. His principal theorem was proven only in the special case that "A is an algebra in which every element, which has no inverse, is nilpotent" (p. 105) and, indeed, is false as stated since a restriction on the field of scalars is necessary. Requiring the field to be of characteristic zero is sufficient to insure the validity of the principal theorem.

Wedderburn's method and results have since come to be seen as revolutionary. Eric Temple Bell, in his address for the semi-centennial of the American Mathematical Society in 1938, observed (p. 30) with reference to the 1907 Wedderburn paper that "algebra, after the assimilation of this paper, was a very different thing from what it was before. Much of it took on the graces of civilized generality and unity." Rather more precisely, Karen Parshall wrote (1982, p. 247):

> Wedderburn . . . revolutionized the theory of algebras with his elegantly simple theory of ideals, his penetrating extension of the Peirce decomposition, and his overall strategy of factoring out the radical and concentrating on the semisimple part of the algebra.

There was, in fact, relatively little work on hypercomplex systems in the years immediately following Wedderburn's 1907 paper. However the question of characterizing division algebras persisted. We recall that Frobenius in 1878 and C. S. Peirce (1881) had shown that the only real linear associative division algebras were the real numbers, the complex numbers, and the quaternions. Wedderburn and Dickson each showed in 1905 that any finite linear associative division algebra is a field. Dickson had described, in 1905 and 1906, division algebras in which one or more of the postulates failed. Again in 1912 Dickson considered division algebras. He showed that any real commutative, non-associative division algebra with identity must have dimension at least six. And in the same paper he showed that in the "octaves" of Graves and Cayley, a non-associative linear algebra of dimension eight over the reals, division is always possible and unique, a fact not reported earlier. In a 1914 paper on "Linear associative algebras and abelian equations" Dickson, citing the structure theorem of Wedderburn to the effect that an algebra over a field is simple if and only if it is the direct product of a matrix algebra and a division algebra, asserted that the problem of division algebras was the chief outstanding problem in the theory of algebras over an arbitrary field. He noted that except for real quaternions the only known associative division algebras are fields. He then exhibited a new class of associative division algebras over an arbitrary field. Wedderburn (1914) used Dickson's results soon thereafter to generate still more division algebras.

It was Dickson who played a major role in determining the direction of most of the research in algebra during the first three decades of this century. During the period 1900 to 1939, while he was at Chicago, he directed the Ph.D. dissertations of sixty-seven students in algebra and number theory. He was also exceedingly prodigious; his bibliography contained 285 items (Albert 1955) at the time of his death, many of them major expositions of a theory. His normal practice seemed to be to work in a specific area, supervise several graduate students in that area, and then write one or more books on the subject. The subject of linear algebras was no exception.

His *Linear Algebras*, one of the Cambridge Tracts in Mathematics and Mathematical Physics, appeared in 1914. An American reviewer (Graustein 1915, p. 511) summed it up:

> A substantial and systematic introduction to general linear algebras, associative and non-associative, a revision of Cartan's theory of linear associative algebras over the field of complex numbers, the results of Wedderburn's theory of such algebras over a general field, the relation of linear algebras to finite and infinite groups and bilinear forms, the consideration of various special algebras and a wealth of historical and bibliographical references in footnotes—in seventy-three pages!

The development was to make no use of the methods or theory of bilinear forms, matrices, or groups even though connections among these concepts were indicated in a brief section. In the preface Dickson indicated some reasons for his choice of development:

> In presenting . . . the main theorems of the general theory, it was necessary to choose between the expositions of Molien, Cartan and Wedderburn (that by Frobenius being based upon bilinear forms and hence outside our plan of treatment). We have not presented the theory of Molien partly because his later proofs depend upon the theory of groups and partly because certain of his earlier proofs have not yet been made correctly by his methods. The more general paper by Wedderburn is based upon a rather abstract calculus of complexes, comparable with the theory of abstract groups. In compensation, he obtains in relatively brief space the main theorems not only for the usual cases of complex and real algebras, but also for algebras the coordinates of whose numbers range over any field.

Thus Dickson's exposition was based on Cartan's work and was confined to the classical case of complex scalars. While he summarized Wedderburn's results (without noting explicitly the need for a restriction on the field of scalars for the validity of Wedderburn's principal theorem) he did not exploit Wedderburn's methods.

In the period from 1914 to 1922 we see work, mainly on the problems of describing or characterizing algebras, by Dickson, Wedderburn, and Dickson's students, especially Olive C. Hazlett (1890–1974). I have already noted papers by Dickson and Wedderburn in 1914 in which they obtained new division algebras. Hazlett, at this point, was interested in the problem of characterizing algebras of low dimension. She wrote both her Master's thesis and her Ph.D. dissertation under Dickson. In her first paper (1914), based on her thesis, she wrote:

> Linear associative algebras of a small number of units, with coordinates in the field of ordinary complex numbers, have been completely tabulated, and their multiplication tables have been reduced to very simple forms. But if we had before us a linear associative algebra, the chances are that its multiplication table would not be in such a form that we could find out readily to what standard form it was equivalent. And so the question arises, "May we not find invariants which completely describe these algebras?"

She then described such algebraic invariants in the case of associative algebras with identity, of dimension two or three, over the complex numbers.

Citing Wedderburn's theorems to the effect that we can characterize the general algebras if we can characterize three special kinds of algebras, namely, simple matrix algebras, division algebras, and nilpotent algebras, she focused on nilpotent algebras in her Ph.D. dissertation and a paper (1916a) following. She found invariants which characterize, in the sense of Dickson (Hazlett 1916a, pp. 115–116), certain nilpotent algebras of small dimension. Other papers of hers (1916b, 1918) and one by MacDuffee (1922) used the theory of algebraic invariants to attack the problem of characterizing algebras.

In 1921 Dickson, in an article on the "Arithmetic of quaternions," initiated a significant change in direction in his study of linear algebras. He noted later (1923, pp. viii–ix) that

> the theory of arithmetics of algebras has been surprisingly slow in its evolution. Quite naturally the arithmetic of quaternions received attention first; the initial theory presented by Lipschitz in his book of 1886 was extremely complicated, while a successful

> theory was first obtained by Hurwitz in his memoir of 1896 (and book of 1919). Du Pasquier, a pupil of Hurwitz, has proposed in numerous memoirs a definition of integral elements of any rational algebra which is either vacuous or leads to insurmountable difficulties . . . . Adopting a new definition, the author develops at length a far-reaching general theory whose richness and simplicity mark it as the proper generalization of the theory of algebraic numbers to the arithmetic of any rational algebra.

In 1923 Dickson's *Algebras and Their Arithmetics* appeared. Although the chief purpose of the book is the development of a general theory of the arithmetics of algebras which furnishes a direct generalization of the classical theory of algebraic numbers, a significant portion of the book is devoted to a presentation of the theory of algebras as it existed at that time. Whereas his *Linear Algebras* of 1914 restricted attention to algebras over the complex numbers (except for a brief summary of some of Wedderburn's structure theorems), this volume develops the theory of algebras over a general field essentially as it was presented by Wedderburn in his 1907 memoir. Wedderburn's principal theorem is stated explicitly and proved for algebras over fields of characteristic zero, however. Dickson also surveyed recent results (Dickson 1914a; Hazlett 1917; Wedderburn 1914, 1921) in the problem of the determination of division algebras.

The reviewer (Hazlett 1924) noted that the "first part of the book is largely Wedderburn's work recast by Dickson, while the second part (on the theory of arithmetics) is entirely due to Dickson." In the major chapter on the theory of arithmetics, Dickson first defined an integer for any associative algebra with identity over the field of *rational* numbers. He then used the results of structure theory in the following way. If an algebra A is not semi-simple then A is the direct sum of a semi-simple subalgebra S of A and the maximal nilpotent invariant subalgebra N of A. Dickson's fundamental result was that the arithmetic of A is known when we know the arithmetic of S. That is, "we may suppress the properly nilpotent elements of an algebra when studying its arithmetic." (p. 187). Similarly, the arithmetic of a semi-simple algebra is known when we know the arithmetic of each of the simple algebras of which it is the direct sum. In turn, Dickson showed that the integral elements of any simple algebra, a direct product of a division algebra and a matrix algebra, are known when those of the division algebra are known. Thus, the problem of arithmetics of algebras is essentially reduced to the case of arithmetics of division algebras. Dickson applied his results to find all integral solutions of various Diophantine equations which had not been completely solved before his work.

Dickson (1928a) presented an outline of the history and main features of his "new branch of number theory" at the International Congress in Toronto in 1924. And at the same meeting he (1928b) and Hazlett (1928), independently, generalized the definition of integer to a notion suitable in an algebra over any algebraic number field, rather than one suitable just in algebras over the rationals. Hazlett elaborated her ideas in a paper in the *Annals* in 1926.

A translation into German of a major revision of Dickson's *Algebras and Their Arithmetics* (with a chapter by A. Speiser on ideal theory) appeared as *Algebren und ihre Zahlentheorie* in 1927. It contained much new material, especially by Dickson on division algebras and by some of Dickson's students on generalized quaternion algebras over the field of rationals. He also noted briefly further investigations in a number of areas; for example, the treatment by Wedderburn of infinite-dimensional algebras in 1924. The chapter on general number theory of algebras was basically rewritten and supplemented with new material. The Dickson–Hazlett definition of integral elements over a general algebraic field was used and the theory developed in this wider setting. The chapter by Speiser treats the theory of ideals in algebras over the rationals. Thus this book summarized much of the work of the American school in algebras and the arithmetic of algebras and made that work accessible to a large audience.

The appearance of Dickson's 1927 book was not the first instance of the reception in Germany of work on algebras by mathematicians in the United States. Van der Waerden recalled in 1975:

> When I came to Göttingen, I took Emmy Noether's course "Gruppentheorie und hyperkomplexe Zahlen" in 1924/25. One of the main subjects in this course was Maclagan Wedderburn's theory of algebras over arbitrary fields. The same subject was treated, in a much improved form, in her course under the same title in 1927/28. . . .[1]

Noether cited Dickson's 1923 and 1927 volumes in her 1929 paper based on notes from that course and referred to Dickson's notion of "crossed product" in her address at the International Congress at Zurich in 1932. Although Noether's remarks suggest that the work of Wedderburn and Dickson did not leave her untouched, it is not the purpose of this paper to investigate the relationship between the work of Noether and that of the Americans.

I will not examine either the influence of the work of mathematicians in the United States on that of Emmy Noether or of her work on them. Rather I have considered just one area of algebra, the theory of hypercomplex number systems, as it developed in the United States between 1870 and 1927, in an effort to understand one aspect of the mathematical context for the reception of Noether's work here.

We saw that the first American contribution was made by Benjamin Peirce whose 1870 address, published posthumously in 1881 as "Linear associative algebra," established the abstract nature of the field, introduced the concepts of idempotent and nilpotent expressions, articulated the Peirce decomposition of an algebra, and attacked the problem of classification and enumeration of algebras. After a period of revision of Peirce's methods the "Chicago school" under the leadership of E. H. Moore and Leonard Eugene Dickson was to become dominant. An emphasis on postulational methods was already apparent in Dickson's 1903 generalization of the definition of linear associative algebras to algebras over arbitrary fields. It was to the Chicago department that J. H. M. Wedderburn came in 1904 for a year in the United States, and it was while he was at Chicago that many of his most significant ideas began to develop. In 1905 he published his result that any finite division algebra is a field. This result was one of several in the effort to describe division algebras. This problem of determining division algebras was seen

[1] The title of the 1924/25 course was "Gruppentheorie"; that of the 1927/28 course was "Hyperkomplexe Grossen und Gruppencharaktere." (Uta C. Merzbach, private communication, January 1982.)

as of even more importance after the structure theory of Wedderburn was articulated in 1907. Although Wedderburn returned to the United States from Scotland in 1909 to assume a position at Princeton, his methods for analyzing the structure of algebras were not readily adopted. Dickson summarized Wedderburn's results in his 1914 volume but based his development on the work of Cartan. In 1923 Dickson opened an essentially new subfield with his book on *Algebras and Their Arithmetics*. His development of algebras was a reworked version of Wedderburn's theory, and his main chapter introduced his generalization of algebraic number theory. The appearance of Dickson's 1927 German revision of his 1923 book marks the end of his major work in algebras and the beginning of his shift almost exclusively towards additive number theory. It is also one signal of a turning point in the nature of the study of algebras, namely the beginning of the confluence of American and German streams of ideas.

The major problems which concerned the researchers in the area of linear algebras were: the characterization and enumeration of algebras of small dimension; the determination of certain algebras over arbitrary fields, especially division algebras; the description of the structure of algebras over arbitrary fields; and finally, the identification of the "integral elements" in linear algebras and generalization of algebraic number theory. From the beginning the work of the Americans was characterized by an attraction for the abstract. We see, too, that virtually all the American algebraists in linear associative algebras worked without recourse to the theory of groups (except as a model), matrix theory, or bilinear forms; that is, they sought to achieve a coherent internal development of the theory.

References

Albert, A. A. 1955. Leonard Eugene Dickson, 1874–1954. *Bulletin of the American Mathematical Society*, **61**: 331–345.

Bell, E. T. 1938. Fifty years of algebra in America, 1888–1938. In *Semicentennial Addresses of the American Mathematical Society*. American Mathematical Society Semicentennial Publications, v. 2, pp. 1–34. New York: American Mathematical Society.

Dickson, Leonard Eugene. 1903a. Definitions of a field by independent postulates. *Transactions of the American Mathematical Society*, **4**: 13–20.

———. 1903b. Definitions of a linear associative algebra by independent postulates. *Transactions of the American Mathematical Society*, **4**: 21–26.

———. 1904. Definitions of a field by independent postulates. *Transactions of the American Mathematical Society*, **5**: 549–550.

———. 1905a. On finite algebras. *Nachrichten von der Gesellschaft der Wissenschaften zu Göttingen*, 358–393.

———. 1905b. On hypercomplex number systems. *Transactions of the American Mathematical Society*, **6**: 344–348.

———. 1906a. Linear algebras in which division is always uniquely possible. *Transactions of the American Mathematical Society*, **7**: 370–390.

———. 1906b. On commutative linear algebras in which division is always uniquely possible. *Transactions of the American Mathematical Society*, **7**: 514–522.

———. 1912. Linear algebras. *Transactions of the American Mathematical Society*, **13**: 59–73.

———. 1914a. Linear associative algebras and abelian equations. *Transactions of the American Mathematical Society*, **15**: 31–46.

———. 1914b. *Linear Algebras*. Cambridge Tracts in Mathematics and Mathematical Physics, No. 16. Cambridge University Press. Reprint. New York: Hafner Publishing Co.

———. 1921. Arithmetic of quaternions. *Proceedings of the London Mathematical Society*, **20**:225–232.

———. 1923. *Algebras and Their Arithmetics*. Chicago: University of Chicago Press. Reprint 1960. New York: Dover Publications.

———. 1927. *Algebren und ihre Zahlentheorie*. (Translation of completely revised and extended manuscript.) Zürich: Orell Füssli.

———. 1928a. Outline of the theory to date of the arithmetics of algebras. In *Proceedings of the International Mathematical Congress held in Toronto, August 11–16, 1924*, ed. J. C. Fields, vol. 1, pp. 95–102. Toronto: The University of Toronto Press.

———. 1928b. Further development of the theory of arithmetics of algebras. In *Proceedings of the International Mathematical Congress held in Toronto, August 11–16, 1924*, ed. J. C. Fields, vol. 1, pp. 173–184. Toronto: The University of Toronto Press.

Epsteen, Saul, and Maclagan-Wedderburn, J. H. 1905. On the structure of hypercomplex number systems. *Transactions of the American Mathematical Society*, **6**:172–178.

Graustein, W. C. 1915. Review of L. E. Dickson, *Linear Algebras*. In *Bulletin of the American Mathematical Society*, **21**:511–522.

Hawkes, H. E. 1902a. Estimate of Peirce's linear associative algebra. *American Journal of Mathematics*, **24**:87–95.

———. 1902b. On hypercomplex number systems. *Transactions of the American Mathematical Society*, **3**:312–330.

Hawkins, Thomas. 1972. Hypercomplex numbers, Lie groups, and the creation of group representation theory. *Archive for History of Exact Sciences*, **8**:243–287.

Hazlett, Olive C. 1914. Invariantive characterization of some linear associative algebras. *Annals of Mathematics*, **16**:1–6.

———. 1916a. On the classification and invariantive characterization of nilpotent algebras. *American Journal of Mathematics*, **38**:109–138.

———. 1916b. On the rational, integral invariants of nilpotent algebras. *Annals of Mathematics*, **18**:81–98.

———. 1917. On the theory of associative division algebras. *Transactions of the American Mathematical Society*, **18**:167–176.

———. 1918. On scalar and vector covariants of linear algebras. *Transactions of the American Mathematical Society*, **19**:408–420.

———. 1924. Two recent books on algebra. *Bulletin of the American Mathematical Society*, **30**:263–270.

———. 1926. The arithmetic of a general algebra. *Annals of Mathematics*, **28**:92–102.

———. 1928. On the arithmetic of a general associative algebra. In *Proceedings of the International Mathematical Congress held in Toronto, August 11–16, 1924*, ed. J. C. Fields, vol. 1, pp. 185–191. Toronto: The University of Toronto Press.

Huntington, Edward V. 1903. Definitions of a field by sets of independent postulates. *Transactions of the American Mathematical Society*, **4**:31–37.

MacDuffee, C. C. 1922. Invariantive characterizations of linear algebras with the associative law not assumed. *Transactions of the American Mathematical Society*, **23**:135–150.

Moore, E. H. 1902a. On the projective axioms of geometry. *Transactions of the American Mathematical Society*, **3**:142–158.

———. 1902b. A definition of abstract groups. *Transactions of the American Mathematical Society*, **3**:485–492.

Parshall, Karen V. H. 1982. The contributions of J. H. M. Wedderburn to the theory of algebras: 1900–10. Ph.D. dissertation. The University of Chicago.

Peirce, Benjamin. 1881. Linear associative algebra. With notes and addenda, by C. S. Peirce, son of the author. *American Journal of Mathematics*, **4**:97–229.

Shaw, James Byrnie. 1907. *Synopsis of Linear Associative Algebra*. Washington, D.C.: Carnegie Institution.

Taber, Henry. 1904. On hypercomplex number systems. *Transactions of the American Mathematical Society*, **5**: 509–548.

Van der Waerden, B. L. 1975. On the sources of my book *Moderne Algebra. Historia Mathematica*, **2**: 31–40.

Wedderburn, J. H. M. 1905. A theorem on finite algebras. *Transactions of the American Mathematical Society*, **6**: 349–352.

———. 1907. On hypercomplex numbers. *Proceedings of the London Mathematical Society*, (2) **6**: 77–118.

———. 1914. A type of primitive algebra. *Transactions of the American Mathematical Society*, **15**: 162–166.

———. 1921. On division algebras. *Transactions of the American Mathematical Society*, **22**: 129–135.

———. 1924. Algebras which do not possess a finite basis. *Transactions of the American Mathematical Society*, **26**: 395–426.

Emmy Noether: Historical Contexts

UTA C. MERZBACH*

"Algebra is fit enough to help us make discoveries, but not fit to bequeath them to posterity." This sentiment, shared by many rigorous seventeenth century mathematicians, aptly characterizes the subservient place algebra held within mathematics for centuries. It is difficult to conceive even the most ardent nineteenth century champion of algebra suggesting that "algebra is the foundation and tool of all mathematics," as Emmy Noether would do in 1931 [1].

In this anniversary year, when we mark the centennial of the births of Emmy Noether and Wedderburn, as well as the sesquicentennial of the death of Galois, it seems appropriate to review briefly some of the factors that went into the transformation of the role of algebra in mathematics—paying particular attention to the place held by Emmy Noether in this process.

It is indeed the work of Galois and his contemporaries that marks a major break in the traditional role and definition of algebra. Up to the 1820s algebra had been identified with the theory of equations. Since the seventeenth century, when Cartesians had viewed it as the key to discovery in mathematics, the chief problem had been that of finding the solution by radicals of the general equation of degree n. With Abel's proof of the insolvability of the quintic, attention shifted from seeking procedures for finding an algebraic solution to either seeking conditions that would determine whether such a solution exists or to finding non-algebraic solutions via analytic or numerical means, as, for example, by elliptic functions or numerical approximation techniques.

The result was an initial fragmentation of problem areas and a gradual interweaving of new concepts. Many problems previously considered as belonging to algebra were now treated as part of function theory, number theory or (algebraic) geometry. Conversely, algebraic investigations came to treat prominently concepts that had either been of secondary importance in the traditional studies of

* Smithsonian Institution, Washington, DC 20560, U.S.A.

equations or were being newly developed. If we look at concepts that were to be of special importance in the work of Emmy Noether, we find a remarkable number evolving from developments in the 1820s and 1830s.

Abel and Galois, in studying conditions for algebraic solutions, based their work on the substitution-theoretic tradition established by Lagrange and the number-theoretic techniques employed by Gauss in his solution of the cyclotomic equation. In their work, the notions of irreducibility, normal invariant subgroup and, implicitly, field extensions all received prominent treatment. Those who cleared the paths marked by them developed the theory of substitution groups and made explicit the concepts of fields and field extensions. At the same time, the notions of invariance and (linear) transformations came to prominence with the development of modern projective geometry in the 1820s. But these same concepts had been significant in number theory, and became especially important as the study of quadratic forms, long of interest to both number theory and algebraic geometry, was placed on a new footing by those who followed the lead of Gauss' *Disquisitiones Arithmeticae*.

After 1827, when Gauss published his classic on differential geometry, the study of differential invariants assumed increasing importance, especially through the work of Riemann in the 1850s. Also following Gauss, specifically his paper of 1832 on complex number theory, were those who developed algebraic number theory, notably the school of Dirichlet; this includes Riemann, Eisenstein, Kummer, Kronecker, and Dedekind. The work of the last three contains the origins of the concepts of ideal, ring, module, prime divisor, and even chain condition. In England, concern with foundations and fascination with the Lagrangian operational calculus of functions led to investigations on the basic operations of numbers, abstracted from traditional areas of study. Following Hamilton's discovery of quaternions in 1843, these concerns led to the creation of the hypercomplex, non-commutative and non-associative systems that Jeanne LaDuke touches upon in her presentation. We must recall these trends going back to the first half of the nineteenth century to understand why we find a greater degree of abstraction in this tradition—why, for example, Wedderburn's early results and the work of the Americans Jeanne LaDuke mentions reflect an abstract point of view at a time when their Continental contemporaries restrict their results to the real and complex fields.

Some trends in the intertwining of concepts mentioned above are illustrated by five works published in 1882, the year of Emmy Noether's birth.

1. Dedekind and Weber's "Theorie der algebraischen Funktionen einer Veraenderlichen" had as the stated aim of the authors placing the theory of algebraic functions of a variable on a new basis free from the geometric reasoning found in Riemann's work. To do this, the authors proceeded in a fashion analogous to that used by Dedekind in his treatment of algebraic numbers in the second and third edition of Dirichlet's lectures on number theory. Specifically, they utilized the concepts of field of algebraic functions and of ideals that permit unique factorization, the algebraic theory of which was used to define parts of a Riemann surface. The Riemann surface defined in this context was then considered invariant with respect to this field.

2. Walther Dyck's "Gruppentheoretische Studien" was the work of a student and assistant of Felix Klein, who had arrived at his study of discrete and continuous groups from Klein's Erlanger Programm. By way of his geometric studies and Klein's influence, he had been led to Cayley, however; to this we can attribute the fact that in his 1882 paper he provided a definition of an abstract group and proceeded to elaborate the concepts of generators of a group, resulting in a study of relations among generators [2].

3. Kronecker's "Grundzuege einer arithmetischen Theorie der algebraischen Groessen" was the product of a man who had been working on number theory and algebraic equation theory since the 1840s. Kronecker motivated this work by his long-standing interest in bringing to the fore the arithmetic viewpoint in algebra and to apply arithmetic methods to specific algebraic questions.

4. Eugen Netto's *Substitutionentheorie und ihre Anwendung auf die Algebra* was written by a student of Kronecker. Netto followed the French textbook writers Serret and Jordan in producing this work, devoted not to the theory of abstract groups but to a synthesis of results in the theory of substitution groups. However, by stressing concepts taken from group-theoretic studies in number theory, and by placing emphasis on the notions of isomorphism and homomorphism, he prepared the way for the various abstract and axiomatic treatments of groups that were to begin flourishing in the 1880s.

5. J. J. Sylvester's "A Word on Nonions," provided an example of an abstract algebraic system characteristic of the British school. In a related note [3], prompted by relevant implications of C. S. Peirce's *Logic of Relatives*, Sylvester commented on the relationship of algebra to logic, observing that "the application of Algebra to Logic is now an old tale—the application of Logic to Algebra marks a far more advanced stadium in the evolution of the human intellect." (Sylvester 1883.)

Though the authors of these five publications proceeded from different points of view and different interest areas, two elements were common to them: One was a degree of abstractness, although barely present in the work of Netto. The other was the attempt to "arithmetize" the subject being treated. This is stressed most explicitly by Dedekind and Weber, as well as by Kronecker, and is consonant with their long-standing efforts to replace analytic proofs in number theory by arithmetic ones. It was this desire to arithmetize number theory and algebra that motivated most of the mathematicians influencing Emmy Noether. Among these, David Hilbert, who at the time of her birth was studying with Heinrich Weber at Koenigsberg, was to be the most systematic exponent of arithmetization and axiomatization.

Emmy Noether was born and raised in Erlangen, in a mathematical environment dominated by invariant theory and algebraic geometry. While Dedekind and Kronecker strove to arithmetize algebraic number theory, Max Noether, the distinguished algebraic geometer who was her father, attempted to "algebraize" algebraic function theory, that is, to free it from analytic or intuitive geometric notions. His long-time colleague and friend at Erlangen, Paul Gordan, specialized in the calculation of invariants, which appeared to be the basic algebraic tool for achieving this purpose. Algebraic invariants had become a subject of study in their own right in England in the 1840s; following the seminal work by George Boole, it was Cayley and Sylvester, along with George Salmon, who pushed the study of algebraic invariants to the forefront of algebraic activity until they could identify "modern algebra" with "invariant theory." (Salmon 1885.) Cayley and Sylvester generated basic invariants and applied them to the solution of equations of degrees 1 to 4. Hilbert, in an 1893 survey of invariant theory repeatedly cited by Emmy Noether, viewed their work as the first stage, the "naive" period of invariant theory; the founders of the so-called "symbolic" method of invariant theory, Clebsch, Aronhold, and Gordan, who initiated the computational phase of invariant theory, Hilbert considered as representing the second, or "formal" phase. As Emmy Noether emphasized, it was Hilbert himself who initiated the third, or critical, phase of the history of that theory with his basis theorem. Hilbert's good friend Minkowski, congratulating him on this achievement which obviated the need to examine practically every form of invariant separately, observed in a letter of 9 February 1892:

> It is really time to level the castle of the highwaymen Stroh, Gordan, Stephanos or whatever their names may be, who overpower invariants traveling singly and incarcerate them in their castle dungeons, ... (Translated from Minkowski 1973:45.)

Emmy Noether's doctoral dissertation of 1907 (Noether 1907, 1908), written under Paul Gordan, still placed her among the highwaymen. Dealing with the construction of the systems of forms for ternary biquadratic forms, it was typical of the computational approach to invariant theory taken by Gordan and his school. It was followed by a study on formal invariants of n-ary forms, presented at the Salzburg meeting of the Deutsche Mathematiker-Vereinigung in 1909 (Noether 1910, 1911). While yet in Erlangen she moved into the Hilbert sphere, however.

Her papers during the next decade show thorough familiarity not only with Hilbert's work on invariant theory and his *Zahlbericht*, but with the classic treatise on field theory of Steinitz which had appeared in the winter of 1909/10.

The problem with which she dealt during her Erlangen period involve two areas: (1) Extensions of Hilbert's results on his basis theorem and partial attacks on Hilbert's Fourteenth Problem; and (2) Study of differential invariants.

The first problem area is characterized by attempts to improve on Kronecker's methods. The major paper in this group is Noether (1915a), "Koerper und Systeme rationaler Funktionen." In this she settled the question of the existence of a finite rational basis for fields of rational functions, and applied the result to special cases of Hilbert's Fourteenth Problem. Further applications followed in Noether (1915b, 1915c). Noether (1915b) furnished an elementary constructive finiteness proof, based on the theory of symmetric functions, for the invariants of a finite group. Noether (1915c) contained proof of a 1914 conjecture by Hilbert, and a proof of an invariant-theoretic theorem on series development found in Mertens. The latter proof utilized results by E. Fischer, Gordan's successor at Erlangen. Noether credited Fischer with stimulating her to pursue most of the work just mentioned; they particularly overlapped in their study of differential invariants.

As her biographers have noted, this work on differential invariants in 1916 brought her to Göttingen, where for several years she assisted Hilbert and Felix Klein in their work on general relativity theory. References in their correspondence and publications provide evidence of the esteem they had for her contributions to the subject (Klein 1921:559–60, 585). In 1918, Felix Klein presented to the Göttingen Society of Sciences two papers by Emmy Noether on differential invariants. The first of these, "Invarianten beliebiger Differentialausdruecke," (Noether 1918b) follows a procedure of proving a "reduction theorem" analogous to that used in Noether (1915c). As she emphasized in the introduction to this paper, she replaced the elimination techniques utilized by Christoffel and Ricci by direct generation of invariants through differentiation processes that included variational processes. Noether (1918c), entitled "Invariante Variationsprobleme," was her "Habilitationsschrift," dedicated to Felix Klein on the fiftieth jubilee of his doctorate. She here dealt with invariance with respect to continuous (Lie) groups. Best known for containing "Noether's Theorem," which is fundamental to the study of correspondences between conservation laws and specific invariances, this paper gained her notice among mathematical physicists. Let us remark that it was not her algebra but this work, specifically (1918c), that brought her to the attention of the American mathematical establishment in the 1920s. More precisely, Oswald Veblen, who published his work on differential invariants as a Cambridge Tract in 1927, initiated a correspondence with her on that subject. In this, Noether observed that she had not had a chance to expand on the paper because she subsequently occupied herself with arithmetical matters (Washington 1927).

The "arithmetical matters" were the beginnings of the algebraic work that we think of when we talk of Emmy Noether, namely her work on ideal theory. It appears to have been as the result of exchanges with Landau that she began to occupy herself intensely with Dedekind after her arrival in Göttingen. If Dedekind's influence hitherto had been less direct—although we should remember that Hilbert and Heinrich Weber stood on his shoulders in much of their number-theoretic work—it now affected her directly and explicitly, and led her to the motto: "Es steht alles schon bei Dedekind." (van der Waerden 1975).

Noether's increased awareness of Dedekind is apparent in several transitional papers appearing between 1916 and 1920. In Noether (1916a) she utilized Dedekind's concept of an isomorphic mapping to examine properties of general abstract fields. It may be noted that here, as well as in Noether (1915d), she first drew heavily on notions of abstract set theory, citing Zermelo primarily. More revealing is her analysis of Dedekind's work in Noether (1919a), the report entitled "Die arithmetische Theorie der algebraischen Funktionen einer Veraenderlichen, in ihrer Beziehung zu den uebrigen Theorien und zu der Zahlkoerpertheorie." In discussing the foundations of the theories of Riemann, Weierstrass, Hensel–Landsberg, Brill–Noether, and Dedekind–Weber, she singled out the last, arithmetic, theory as preserving "full purity of method." Noether (1918a) dealt with polynomials having a given Galois group. It was her first effort at resolving the question of the existence of a Galois extension of a given field with Galois group isomorphic to a given group [4]. In Noether–Schmeidler (1920A), motivated by generalizing results concerning the decomposition of linear differential expressions to those in n variables, the authors developed a theory of modules composed of polynomials with non-commutative multiplication. This involved introduction of the concepts of "left" and "right" module. The approach of Dedekind's Eleventh Supplement prevails in this paper, which has been called "a sort of prologue to her general theory of ideals," an opinion apparently shared by herself (Alexandroff 1936).

Her general theory of ideals was put forth in two classical papers (Noether 1921a, 1926a). She described the first, "Idealtheorie in Ringbereichen," as dealing with the transfer of decomposition theorems for ideals in algebraic number fields to those for ideals in arbitrary rings. She recast the traditional formulation for unique factorization by singling out four characteristic properties of the prime-power factors of an integer: those of having no common divisors; of being relatively prime; of being primary; and of being irreducible. Positing a finiteness condition on the general (commutative) ring under consideration, she then showed that there are generally four decompositions, corresponding to the four characteristic properties. She showed that rings meeting the finiteness condition satisfy the equivalent of Dedekind's ascending chain condition. In the proofs that followed she provided an impressive demonstration of the power gained from the use of the chain condition. She also noted to what extent the results concerning irreducible ideals can be carried over to non-commutative rings. In the introduction to the paper, Noether commented with customary care on the extent to which some of the results she was unifying had been obtained by others: Lasker and Macaulay had treated the case of primary ideals for polynomials over the complex field using elimination theory; Schmeidler had done the same for classes of ideals without common divisors. Noether elaborated on the extent to which her joint 1920 publication with Schmeidler sharpened his earlier results, and how both of the earlier publications differed from the present one. Finally, she noted the more abstract work of Fraenkel, who, in several papers since 1914, had treated ring decompositions; Fraenkel had also provided an axiomatic treatment of rings in his "Habilitationsschrift" of 1916.

In Noether (1926a), "Abstrakter Aufbau der Idealtheorie in algebraischen Zahl- und Funktionenkoerpern," she aimed at an abstract characterization of rings whose ideals can be represented as products of the powers of prime ideals. Aside from its content, this paper is particularly striking for the readability it has for the modern mathematician, and the care with which Noether relates the concepts and results to those of Dedekind from which they are derived [5].

While the two preceding papers may be considered the cornerstones of her general ideal theory, Noether published several related papers during this period. Aside from the publication of specific results (Noether 1922a, 1923d, 1925a) we may note a series of papers (Noether 1921c, 1921d, 1923b, 1923c, 1924a) dealing with theories of elimination, of zeros of polynomials, and of ideals; as noted in Noether (1924a), she strove to show a parallelism in elimination theory and the theory of zeros of polynomials by relating Hentzelt's elimination theory, Steinitz's field theory and her own ideal theory.

It was in the mid 1920s, after she had received her appointment as "nichtbeamteter ausserordentlicher Professor" and her "Lehrauftrag" for algebra, that she turned her attention to the remaining algebraic tradition: the study of hypercomplex systems and group representations. In lectures on group theory delivered in 1924/25 and in a paper read at the 1925 meeting of the Deutsche Mathematiker-Vereinigung in Danzig she first outlined the treatment of finite group representations from an ideal-theoretic point of view: "The Frobenius theory of group characters ... is interpreted as the ideal theory of a completely reducible ring, the group ring." (Noether 1925c). This work was expanded during the next few years, receiving a rounding off in her 1927/28 lectures on "Hypercomplex quantities and group characters"; van der Waerden's lectures notes for these formed the basis for Noether (1929a). The reader of this publication immediately notes the extent to which Noether had absorbed not only the hypercomplex treatments of Frobenius and Molien, the structural analysis of Dickson and Wedderburn, but also the direct representation theory of Burnside and Schur.

Also belonging to this period (Noether 1926b). "Der Endlichkeitssatz der Invarianten endlicher linearer Gruppen der Charakteristik p," in which she returned to a former problem area, provides a particularly clear example of increased sharpness and generality of results obtainable by newly available concepts. Noether 1927a, in which she obtained a generalization of Dedekind's theorem on discriminants, is treated as the first step leading to a general ramification theory. She took further steps in (1932a), "Normalbasis bei Koerpern ohne hoehere Verzweigung," where she analyzed relationships between the existence of normal bases and ramification of field extensions [6].

Overlapping with these investigations is a series of papers leading to the abstract treatment of class field theory. She collaborated with Richard Brauer (1927A) in a study of minimal splitting fields, which led to new insights concerning Schur's theory of group representations [7]. She then applied the theory of splitting fields to those of crossed products and of factor systems. As noted in Noether (1932b, 1933a), she "smelted" Dickson's notion of crossed product with Brauer's theory of factor systems, derived from the tradition of Schur and Speiser, in her lectures of 1929/30. This, she explained, provided her with the proper tool for attacking problems in commutative algebra by means of noncommutative systems. She formulated this program clearly in her address to the 1932 International Congress at Zürich:

> By means of the theory of algebras one seeks to gain invariant and simple formulations for known facts concerning quadratic forms or cyclic fields, i.e. those formulations, that depend only on the structural properties of the algebras. Once these invariant formulations have been proved . . . one has won a transfer of these facts to arbitrary Galois fields. (Noether 1932b).

After noting the concurrent work being done by Hasse, Chevalley, and Artin she proceeded to illustrate her program by dealing with the principal genus theorem and the norm principle. This useful strategy permitted her to call attention to the "Haupsatz" that any "normal" division algebra over an algebraic number field is cyclic, which had been proved by three reductions (Noether, Brauer, and Hasse 1932A): to show how the norm principle proved analytically in class field theory can be proved algebraic-arithmetically by considering a cyclic splitting field and a base field which is an algebraic number field; and, in discussing the principal genus theorem, to indirectly suggest the potential importance of further development of ramification theory and of a theory of Galois modules. The subject was treated in more generality in Noether (1933a); the translation of the principal genus theorem to the noncommutative case in (1933b); questions for further consideration were raised in (1934). It should be noted that class field theory was the subject of her last guest lectures at Princeton.

A review of the change in the place of algebra to which Emmy Noether contributed must include mention of the "algebraization" of topology for which Alexandroff and Hopf gave her full credit in the preface to their classical *Topologie* (1935). Relevant in this connection is a letter of January 1929 which Paul Alexandroff, recently returned to Moscow after a stay in Princeton, wrote to Oswald Veblen. After describing his own heavy teaching load at Moscow and Smolensk—he was carrying a course and seminar in topology in addition to work in analysis and two courses in algebra—he added:

> This winter (as you know) we have Miss Noether here in Moscow as guest professor, and of course her presence enlivens our mathematical life greatly, particularly since algebra notably belongs to those mathematical fields that unfortunately have been cultivated little so far in Moscow. Partly under the influence of Miss Noether, partly also stimulated by my algebraic lectures here in Smolensk, I begin to become very interested in algebra, for the time being only "from afar" of course, without trying to work in it myself. The topological lecture which Hopf is holding in Berlin this winter (and which is very interesting . . .) is also very strongly algebraically influenced; thus f.i. he handles the entire theory of topological invariants based on "homology concepts" purely group theoretically, and I find that very beautiful: any calculation is brought to a minimum thereby . . . Since I have many German friends, I have even placed a large, comfortable armchair in my room (there is not enough room for a sofa!) and Miss Noether sits in that armchair when she visits me! (Translated from Washington 1929.)

Alexandroff's armchair would not long be used by those for whom it was intended. But the concepts brought to the fore by Emmy Noether and her associates of the late 1920s were to become part of the armamentarium of mathematicians in Göttingen and Nancy as well as Moscow and Princeton. The remarkable change they induced in the language and methods of mathematics is apparent upon contrasting nearly any early twentieth century mathematical publication with a contemporary one.

In crediting Emmy Noether with her share in this transformation of mathematics, most biographers have followed Herman Weyl's analysis of her work, noting that it falls into three periods, of which the first, lasting until about 1919, was one of "relative dependence," whereas the other two were characterized by the algebraic work for which she is remembered. It becomes apparent that difficulties arise in drawing so sharp a distinction between that work which is "relatively dependent" and the rest, however. One can find examples of originality in her early work, and many instances of dependence in the productions of her later period—although that is not a particularly fruitful exercise. More importantly, the exclusion of "dependent" work from consideration makes it impossible to study any process of conceptual change. It is natural for us to pay most attention to the post-1919 period of her work. The work that was most influential was done when she was in her forties; The "Noether school" of those who collaborated with her in attempting to make algebra the tool and foundation of all of mathematics consists of individuals who knew her only during the last decade of her life. In short, her historic influence in effecting conceptual change is based on the events of the last decade of her life [8]. Her stature as a creative mathematician is better understood if we examine her mathematical career in its entirety, however. Only then can we appreciate to what extent Emmy Noether's work fits Poincare's famous description of mathematical creativity:

> To invent is to discern, it is to choose. . . . Mathematical facts worthy of being studied are those which by their analogy with other facts lead us to the knowledge of a mathematical law in the same fashion that experimental facts lead us to the knowledge of a physical law. They are those which reveal to us unsuspected relations between other facts, known for a long time, but wrongly believed to be strangers to one another. Among the combinations one chooses the most fruitful are often those which are formed of elements borrowed from widely separated domains. (Poincaré 1908:48–49.)

Noether was well-versed in the mathematical literature of her time and in major works of the second half of the nineteenth century. Her publications show how well she absorbed the conceptual developments of her predecessors. The more than 70 reviews she wrote provide examples of the perceptive analyses to which she subjected the contributions of her contemporaries. Her deep knowledge of the literature and her ability to recognize and bring to the fore those concepts that would prove most fruitful prepared her to choose the combinations that allowed her to undertake her grand synthesis.

If one examines her work after 1910, one finds continual growth, but little change in methodological pattern. Van der Waerden noted that "the module-theoretic concepts of direct sum and intersection decomposition, residue class modules, and module isomorphism," which are utilized in Noether and Schmeidler (1920A), "run like red threads through her later work." (van der Waerden 1935.) One may add that the methodological concepts of arithmetization, generalization, abstraction, reduction, and transfer are the spindles she used to trim and combine in an orderly fashion the algebraic threads that had been generated, separated, and entangled with geometric and analytic strands during the preceding century [9].

Notes

[1] The full statement "nicht eigentlich eine Disziplin fuer sich, sondern Grundlage und Werkzeug der gesamten Mathematik" appears in her review of volume 5 of Kronecker's *Werke* (Noether 1931R2).

In the following, references to Noether's publications relate to the bibliography at the end of this volume.

[2] A more immediately influential paper by H. Weber (Weber 1882) on the abstract theory of groups appeared in the same volume of the *Mathematische Annalen*.

[3] This is appended to Sylvester (1882) in the *Collected Mathematical Papers*.

[4] Noether (1918a) is one of several papers discussed in Smith (1976:2–3); also see Swan (1981).

[5] For an extensive discussion of Noether (1921a, 1924b) see Gilmer (1981).

[6] See the analysis in Smith (1976) and Froehlich (1981).

[7] Also see Lam (1981)

[8] Ironically, it was probably the very fact that she never received a regular appointment at Göttingen which allowed her to organize her algebraic research as single-mindedly as she did, to

display that generosity towards her followers to which van der Waerden, Alexandroff, and others have given eloquent testimony, and to engage so fully in the editing of Dedekind's works and selected correspondence.

[9] Readers unfamiliar with nineteenth century mathematical terminology may wish to refer to Kline (1972) for definitions of several terms used in the foregoing. The same work provides bibliographic references to the classical works referred to above. Noether's papers carry full bibliographic references in most instances.

References

Alexandroff, Paul. 1936. *Uspekhi Mat. Nauk* **2**: 254–265. Translated as "In Memory of Emmy Noether" in Brewer and Smith 1981.

Alexandroff, Paul and Hopf, Heinz. 1935. *Topologie*. Berlin: Julius Springer.

Brewer, James W. and Smith, Martha K. 1981. *Emmy Noether. A Tribute to Her Life and Work*. New York and Basel: Marcel Dekker.

Dedekind, Richard. 1969. *Gesammelte mathematische Werke*. 3 vols. Bronx, N.Y.: Chelsea. (= reprint of 1930–1932 edition.)

Dedekind, R. & Weber, H. 1882. "Theorie der algebraischen Funktionen einer Veraenderlichen." *Journal fuer die reine und angewandte Mathematik* **92**: 181–290.

Dick, Auguste. 1970. *Emmy Noether 1882–1935*. Basel: Birkhaeuser.

Dyck, Walter 1882. "Gruppentheoretische Studien." *Mathematische Annalen* **20**: 1–44.

Froehlich, A. 1981. "Algebraic Number Theory." Pp. 157–163 in Brewer and Smith 1981.

Gilmer, Robert 1981. "Commutative Ring Theory." Pp. 131–143 in Brewer and Smith 1981.

Hilbert, David. 1890. "Ueber die Theorie der algebraischen Formen." *Mathematische Annalen* **36**: 473–534 (= *Gesammelte Abhandlungen* **2**: 199–257).

———. 1893. "Ueber die vollen Invariantensysteme." *Mathematische Annalen* **42**: 313–373 (= *Gesammelte Abhandlungen* **2**: 287–344).

———. 1896. "Ueber die Theorie der algebraischen Invarianten." *Mathematical Papers Read at the International Mathematical Congress Chicago 1893*. Pp. 116–124. New York: Macmillan & Co. (= *Gesammelte Abhandlungen* **2**: 376–383).

———. 1897. "Die Theorie der algebraischen Zahlkoerper." *Jahresbericht der Deutschen Mathematiker-Vereinigung* **4**: 175–546 (= *Gesammelte Abhandlungen* **1**: 163–363).

———. 1914a. "Ueber die Invarianten eines Systems von beliebig vielen Grundformen." Pp. 448–451 in *Mathematische Abhandlungen Hermann Amandus Schwarz zu seinem fuenfzigsten Doktorjubilaeum*. Berlin: Julius Springer. (= *Gesammelte Abhandlungen* **2**: 390–392.)

Kimberling, Clark. 1981. "Emmy Noether and her Influence." Pp. 3–61 in Brewer and Smith 1981.

Klein, Felix. 1921. *Gesammelte Mathematische Abhandlungen* 1. Berlin: Julius Springer.

Kline, Morris. 1972. *Mathematical Thought from Ancient to Modern Times*. New York: Oxford University Press.

Kronecker, L. 1882. "Grundzuege einer arithmetischen Theorie der algebraischen Groessen." *Journal fuer die reine und angewandte Mathematik* **92**: 1–122 (= *Werke* **2**: 237–387).

Lam, T. Y. 1981. "Representation Theory." Pp. 145–155 in Brewer and Smith 1981.

Minkowski, Hermann. 1973. *Briefe an David Hilbert*. Edited by L. Ruedenberg and H. Zassenhaus. Berlin, Heidelberg, New York: Springer-Verlag.

Netto, Eugen. 1882a. *Substitutionentheorie und ihre Anwendung auf die Algebra*. Leipzig: B. G. Teubner.

Poincare, Henri. 1908. *Science et Methode*. Paris: Flammarion.

Salmon, George. 1859. *Lessons Introductory to the Modern Higher Algebra*. Dublin: Hodges, Smith & Co. (4th edition = 1885).

Smith, Martha K. 1976. "Emmy Noether's Contribution to Mathematics." 13 pp.

Steinitz, Ernst. 1909–1910. "Algebraische Theorie der Koerper." *Journal fuer die reine und angewandte Mathematik* **137**: 167–246 and 247–309.

Swan, Richard G. 1981. "Galois Theory." Pp. 115–124 in Brewer and Smith 1981.

Sylvester, J. J. 1882. "A Word on Nonions." *Johns Hopkins University Circulars* **1**: 241–242. (= *Collected Mathematical Papers* **3**: 647–649).

———. 1883. Note on the preceding. *Johns Hopkins University Circulars* **2**: 46. (= *Collected Mathematical Papers* **3**: 649–650).

Veblen, O. 1927. *Invariants of Quadratic Differential Forms*. Cambridge Tracts in Mathematics and Mathematical Physics . . . no. 24. Cambridge: University Press.

Van der Waerden, B. L. 1935. "Nachruf auf Emmy Noether." *Mathematische Annalen* **111**: 469–474. (Reprinted in Dick 1970; translated in Brewer and Smith 1981.)

Washington, D. C. 1927 & 1929. The Library of Congress. Oswald Veblen Papers. General Correspondence. (Letters from Paul Alexandroff of 8 January 1929 and from Emmy Noether of 29 September 1927.)

Weyl, Hermann. 1935. "Emmy Noether." *Scripta Mathematics* **3**: 201–220. (Reprinted in Dick 1970.)

Bibliography*

I. Articles, Abstracts of Presented Papers, Editions of Work by Others

1907. "Über die Bildung des Formensystems der ternären biquadratischen Form." *Sitzungsberichte der Physikalisch-Medizinischen Sozietät* 39: 176–179.
excerpt from introduction and chapter 1 of dissertation

1908. "Über die Bildung des Formensystems der ternären biquadratischen Form." *Journal für die reine und angewandte Mathematik* 134: 23–90 + 1 table.
dissertation

1910. "Zur Invariantentheorie der Formen von n Variabeln." *Jahresbericht der Deutschen Mathematiker-Vereinigung* 19: 101–104.
presented at the Salzburg DMV meeting in 1909

1911. "Zur Invariantentheorie der Formen von n Variabeln." *Journal für die reine und angewandte Mathematik* 139: 118–154.
expanded publication of preceding

1914. "Rationale Funktionenkörper." *Jahresbericht der Deutschen Mathematiker-Vereinigung* 22: 316–319.
presented at the Vienna DMV meeting in 1913

1915a. "Körper und Systeme rationaler Funktionen." *Mathematische Annalen* 76: 161–196.
dated Erlangen, May 1914

1915b. "Der Endlichkeitssatz der Invarianten endlicher Gruppen." *Mathematische Annalen* 77: 89–92.
dated Erlangen, May 1915

1915c. "Über ganze rationale Darstellung der Invarianten eines Systems von beliebig vielen Grundformen." *Mathematische Annalen* 77: 93–102.
dated Erlangen, 5 January 1915

1915d. "Die allgemeinsten Bereiche aus ganzen transzendenten Zahlen." *Mathematische Annalen* 77: 103–128.
dated Erlangen, 30 March 1915

1915e. "Endlichkeitsfragen der Invariantentheorie." *Jahresbericht der Deutschen Mathematiker-Vereinigung* 24 pt. 2: 68.
title of paper presented to the Mathematische Gesellschaft in Göttingen on 13 July 1915

1916a. "Die Funktionalgleichungen der isomorphen Abbildung." *Mathematische Annalen* 77: 536–545.
dated Erlangen, 30 October 1915

* This complete Bibliography of Emmy Noether's publications was supplied by Uta Merzbach—Editors.

1916b. "Über ganze transzendente Zahlen." *Jahresbericht der Deutschen Mathematiker-Vereinigung* 24 pt. 2:111.
title of paper presented to the Mathematische Gesellschaft in Göttingen on 9 November 1915

1916c. "Alternative bei nichtlinearen Gleichungssystemen." *Jahresbericht der Deutschen Mathematiker-Vereinigung* 25 pt. 2:31.
title of paper presented to the Mathematische Gesellschaft in Göttingen on 1 February 1916

1917. "Laskers Zerlegungssatz der Modultheorie." *Jahresbericht der Deutschen Mathematiker-Vereinigung* 26 pt. 2:31.
title of paper presented to the Mathematische Gesellschaft in Göttingen on 19 June 1917

1918a. "Gleichungen mit vorgeschriebener Gruppe." *Mathematische Annalen* 78:221–229.
dated Göttingen, July 1916;
paper of same title presented to the Mathematische Gesellschaft in Göttingen on 23 May 1916

1918b. "Invarianten beliebiger Differentialausdrücke." *Nachrichten von der Gesellschaft der Wissenschaften zu Göttingen* 37–44.
presented to the Society by Felix Klein on 25 January 1918;
paper of same title presented to the Mathematische Gesellschaft in Göttingen on 15 January 1918

1918c. "Invariante Variationsprobleme." *Nachrichten von der Gesellschaft der Wissenschaften zu Göttingen* 235–257.
presented to the Society by Felix Klein 26 July 1918;
final printed version submitted September 1918

1919a. "Die arithmetische Theorie der algebraischen Funktionen einer Veränderlichen in ihrer Beziehung zu den übrigen Theorien und zu der Zahlkörpertheorie." *Jahresbericht der Deutschen Mathematiker-Vereinigung* 28:182–203.
presentation concerning this report made to the Mathematische Gesellschaft in Göttingen on 1 June 1920

1919b. "Die Endlichkeit des Systems der ganzzahligen Invarianten binärer Formen." *Nachrichten von der Gesellschaft der Wissenschaften zu Göttingen* 138–156.
presented to the Society by Felix Klein 27 March 1919

1919c. "Endlichkeit ganzzahliger binärer Invarianten." *Jahresbericht der Deutschen Mathematiker-Vereinigung* 28 pt. 2:29.
abstract of paper presented to the Mathematische Gesellschaft in Göttingen on 5 November 1918

1919d. "Über ganzzahlige Polynome und Potenzreihen." *Jahresbericht der Deutschen Mathematiker-Vereiningung* 28 pt. 2:29–30.
abstract of paper presented to the Mathematische Gesellschaft in Göttingen on 26 November 1918

1920a. "Zur Reihenentwicklung in der Formentheorie." *Mathematische Annalen* 81:25–30.
received by *MA* September 1919

1920b. "Zerlegungssätze der Modultheorie." *Jahresbericht der Deutschen Mathematiker-Vereinigung* 29 pt. 2:54.
abstract of paper presented to the Mathematische Gesellschaft in Göttingen on 10 February 1920

1920c. "Fragen der Modul- und Idealtheorie." *Jahresbericht der Deutschen Mathematiker-Vereinigung* 29 pt. 2:46.
title of paper presented at the Bad Nauheim DMV meeting on 24 September 1920

1920A. (with W. Schmeidler). "Moduln in nichtkommutativen Bereichen, insbesondere aus Differential- und Differenzenausdrücken." *Mathematische Zeitschrift* 8:1–35.
dated Göttingen, 1 August 1919;
received by *MZ* 4 August 1919

1921a. "Idealtheorie in Ringbereichen." *Mathematische Annalen* 83:24–66.
dated Erlangen, October 1920;
received by *MA* 16 October 1920

1921b. "Elementarteiler und allgemeine Idealtheorie." *Jahresbericht der Deutschen Mathematiker-Vereinigung* 30 pt. 2:32.
abstract of paper presented to the Mathematische Gesellschaft in Göttingen on 25 January 1921

1921c. "Über eine Arbeit des im Krieg gefallenen K. Hentzelt zur Eliminationstheorie." *Jahresbericht der Deutschen Mathematiker-Vereinigung* 30 pt. 2:48.
abstract of paper presented to the Mathematische Gesellschaft in Göttingen on 5 July 1921

1921d. "Über eine Arbeit des im Kriege gefallenen K. Hentzelt zur Eliminationstheorie." *Jahresbericht der Deutschen Mathematiker-Vereinigung* 30 pt. 2:101.
summary of paper presented at the Jena DMV meeting on 23 September 1921

1921e. "(Über eine Formel von Lipschitz)." *Jahresbericht der Deutschen Mathematiker-Vereinigung* 30 pt. 2:48.
comment made at meeting of the Mathematische Gesellschaft in Göttingen on 26 July 1921

1922a. "Ein algebraisches Kriterium für absolute Irreduzibilität." *Mathematische Annalen* 85:26–33.
dated Erlangen, August 1921;
received by *MA* 28 August 1921

1922b. "Formale Variationsrechnung und Differentialinvarianten." In: 'Differentialinvarianten' by R. Weitzenböck. *Encyklopaedie der mathematischen Wissenschaften* III, 3:68–71.

1923a. "Algebraische und Differentialinvarianten." *Jahresbericht der Deutschen Mathematiker-Vereinigung* 32:177–184.
report given at the Leipzig DMV meeting on 18 September 1922;
received by *JDMV* on 26 October 1922

1923b. edition of K. Hentzelt's "Zur Theorie der Polynomideale und Resultanten." *Mathematische Annalen* 88:53–79.
received by *MA* 17 March 1922

1923c. "Eliminationstheorie und allgemeine Idealtheorie." *Mathematische Annalen* 90:229–261.
dated Göttingen, 8 May 1923;
received by *MA* 9 March 1923 (sic!)

1923d. "Das Analogon eines Hilbertschen Satzes in der allgemeinen Idealtheorie." *Jahresbericht der Deutschen Mathematiker-Vereinigung* 32 pt. 2:20–21.
abstract of paper presented to the Mathematische Gesellschaft in Göttingen on 7 November 1922

1923e. "Gleichungen mit vorgeschriebener Gruppe." *Jahrbuch über die Fortschritte der Mathematik* 46:135.
abstract of 1918a

1923f. "Die Funktionalgleichungen der isomorphen Abbildung." *Jahrbuch über die Fortschritte der Mathematik* 46:170–171.
abstract of 1916

1923g. "Invarianten beliebiger Differentialausdrücke." *Jahrbuch über die Fortschritte der Mathematik* 46:675.
abstract of 1918b

1923h. "Invariante Variationsprobleme." *Jahrbuch über die Fortschritte der Mathematik* 46:770.
abstract of 1918c

1923i. "Koerper und Systeme rationaler Funktionen." *Jahrbuch über die Fortschritte der Mathematik* 46:1442–43.
abstract of 1915a

1924a. "Eliminationstheorie und Idealtheorie." *Jahresbericht der Deutschen Mathematiker-Vereinigung* 33:116–120.
abstract of paper presented at the Marburg DMV meeting on 25 September 1923;
received by *JDMV* on 19 October 1923

1924b. "Abstrakter Aufbau der Idealtheorie im algebraischen Zahlkörper." *Jahresbericht der Deutschen Mathematiker-Vereinigung* 33 pt. 2:102.
presented at the Innsbruck DMV meeting on 25 September 1924

1924c. "Galoissche Theorie in beliebigen Körpern." *Jahresbericht der Deutschen Mathematiker-Vereinigung* 33 pt. 2:119.
abstract of discussion at meeting of the Mathematische Gesellschaft in Göttingen on 27 November 1923

1924d. "Idealtheorie im algebraischen Zahlkörper." *Jahresbericht der Deutschen Mathematiker-Vereinigung* 33 pt. 2:120.
abstract of paper presented to the Mathematische Gesellschaft in Göttingen on 29 July 1924

1924e. "Zur Reihenentwicklung in der Formentheorie." *Jahrbuch über die Fortschritte der Mathematik* 47:89.
abstract of 1920a

1924f. "Moduln in nichtkommutativen Bereichen, insbesondere aus Differential- und Differenzenausdrücken." *Jahrbuch über die Fortschritte der Mathematik* 47: 97–98.
abstract of 1920A

1924g. "Die arithmetische Theorie der algebraischen Funktionen ..." *Jahrbuch über die Fortschritte der Mathematik* 47:349–350.
abstract of 1919a

1925a. "Hilbertsche Anzahlen in der Idealtheorie." *Jahresbericht der Deutschen Mathematiker-Vereinigung* 34 pt. 2:101.
abstract of paper presented to the Mathematische Gesellschaft in Göttingen on 11 November 1924

1925b. "Ableitung der Elementarteilertheorie aus der Gruppentheorie." *Jahresbericht der Deutschen Mathematiker-Vereinigung* 34 pt. 2:104.
abstract of paper presented to the Mathematische Gesellschaft in Göttingen on 27 January 1925

1925c. "Gruppencharaktere und Idealtheorie." *Jahresbericht der Deutschen Mathematiker-Vereinigung* 34 pt. 2:144.
abstract of paper presented to the Danzig DMV meeting on 15 September 1925

1925d. "Idealtheorie in Ringbereichen." *Jahrbuch über die Fortschritte der Mathematik* 48:121–122.
abstract of 1921a

1926a. "Abstrakter Aufbau der Idealtheorie in algebraischen Zahl- und Funktionenkörpern." *Mathematische Annalen* 96:26–61.
received by *MA* on 13 August 1925

1926b. "Der Endlichkeitssatz der Invarianten endlicher linearer Gruppen der Charakteristik p." *Nachrichten von der Gesellschaft der Wissenschaften zu Göttingen* 28–35.
presented to the Society by Richard Courant on 30 July 1926

1926c. "Der Dedekindsche Diskriminantensatz für beliebige Ordnungen." *Jahresbericht der Deutschen Mathematiker-Vereinigung* 35 pt. 2:125.
title of paper presented to the Mathematische Gesellschaft in Göttingen on 1 December 1925

1926d. "Artins Untersuchungen über Problem 17 von Hilbert (Zerlegung definiter Funktionen in Quadrate)." *Jahresbericht der Deutschen Mathematiker-Vereinigung* 35 pt. 2:125.
title of paper presented to the Mathematische Gesellschaft in Göttingen on 27 July 1926

1927a. "Der Diskriminantensatz für die Ordnungen eines algebraischen Zahl- oder Funktionenkörpers." *Journal für die reine und angewandte Mathematik* 157:82–104.
dated Göttingen, 30 March 1926

1927b. "Algebraische und Differentialinvarianten." *Jahrbuch über die Fortschritte der Mathematik* 49:68.
abstract of 1923a

1927A. (with Richard Brauer). "Ueber minimale Zerfällungskörper irreduzibler Darstellungen." *Sitzungsberichte der Preussischen Akademie der Wissenschaften* 221–228.

1927B. (with H. Kapferer). "(Zusatz zu) Notwendige und hinreichende Multiplizitätsbedingungen zum Noetherschen Fundamentalsatz der algebraischen Funktionen." *Mathematische Annalen* 97:559–567.
abstract of related discussion appeared in *Jahresbericht der Deutschen Mathematiker-Vereinigung* 36 pt. 2:21.

1928a. "Hyperkomplexe Grössen und Darstellungstheorie in arithmetischer Auffassung." *Atti Congresso Bologna* 2:71–73.
summary of presentation to section at IMC Bologna September 1928

1929a. "Hyperkomplexe Grössen und Darstellungstheorie." *Mathematische Zeitschrift* 30:641–692.
received by *MZ* on 12 August 1928

1929b. "Über Maximalbereiche aus ganzzahligen Funktionen." *Rec. Soc. Math. Moscow* 36:65–72.

1929c. "Differentialquotienten von Idealen und Verzweigungstheorie." *Jahresbericht der Deutschen Mathematiker-Vereinigung* 38 pt. 2:81.
title of paper presented to the Mathematische Gesellschaft in Göttingen on 7 February 1928

1929d. "Quaternionenkörper." *Jahresbericht der Deutschen Mathematiker-Vereinigung* 38 pt. 2:81.
title of paper presented to the Mathematische Gesellschaft in Göttingen on 22 November 1927

1929e. "Differentialquotienten von Idealen und Verzweigungstheorie." *Jahresbericht der Deutschen Mathematiker-Vereinigung* 38 pt. 2:81.
title of paper presented to the Mathematische Gesellschaft in Göttingen on 7 February 1928

1929f. "Reisebericht." *Jahresbericht der Deutschen Mathematiker-Vereinigung* 38 pt. 2:142.
title of paper presented to the Mathematische Gesellschaft in Göttingen on 4 June 1929

1929g. "Differentialsätze." *Jahresbericht der Deutschen Mathematiker-Vereinigung* 38 pt. 2:142.
title of paper presented to the Mathematische Gesellschaft in Göttingen on 2 July 1929

1930a. "Idealdifferentiation und Differente." *Jahresbericht der Deutschen Mathematiker-Vereinigung* 39 pt. 2:17.
abstract of paper presented at Prague DMV meeting in 1929

1930b. "Notwendige und hinreichende Multiplizitätsbedingungen ...." *Jahrbuch über die Fortschritte der Mathematik* 53:342.
abstract of 1927B

1930A. (ed., with R. Fricke and Ö. Ore). *R. Dedekind. Gesammelte mathematische Werke.* Vol. 1. Braunschweig: F. Vieweg & Sohn, A.-G.

1931a. "Hyperkomplexe Struktursätze mit zahlentheoretischen Anwendungen." *Jahresbericht der Deutschen Mathematiker-Vereinigung* 41 pt. 2:16.
title of paper presented at the Mathematical Colloquium of the University of Marburg in a lecture series on hypercomplex systems, on 27 February 1931

1931A. (ed., with R. Fricke and Ö. Ore). *R. Dedekind. Gesammelte mathematische Werke.* Vol. 2. Braunschweig: F. Vieweg & Sohn.

1932a. "Normalbasis bei Körpern ohne höhere Verzweigung." *Journal für die reine und angewandte Mathematik* 167:399–404.
received by *JraM* on 24 August 1931

1932b. "Hyperkomplexe Systeme in ihren Beziehungen zur kommutativen Algebra und Zahlentheorie." *Verhandlungen Intern. Math.-Kongress Zuerich* 1:189–194.
presented in plenary session of IMC Zuerich September 1932

1932c. "Hyperkomplexe Systeme und zahlentheoretische Anwendungen." *Jahresbericht der Deutschen Mathematiker-Vereinigung* 42 pt. 2:89.
title of paper presented to the Mathematische Gesellschaft in Göttingen on 19 May 1931

1932d. "Kleinsches Formenproblem und Galoissche Theorie in hyperkomplexer Auffassung." *Jahresbericht der Deutschen Mathematiker-Vereinigung* 42 pt. 2:89.
title of paper presented to the Mathematische Gesellschaft in Göttingen on 19 January 1932

1932A. (with R. Brauer and H. Hasse). "Beweis eines Hauptsatzes in der Theorie der Algebra." *Journal für die reine und angewandte Mathematik* 167:399–404.
received by *JraM* 11 November 1931

1932B. (ed., with R. Fricke and Ö. Ore). *R. Dedekind. Gesammelte mathematische Werke*. Vol. 3. Braunschweig: F. Vieweg & Sohn, A.-G.

1933a. "Nichtkommutative Algebra." *Mathematische Zeitschrift* 37:514–541.
received by *MZ* 8 June 1932

1933b. "Der Hauptgeschlechtssatz für relativ-galoissche Zahlkörper." *Mathematische Annalen* 108:411–419.
received by *MA* 27 October 1932

1933c. "Beziehungen zwischen hyperkomplexer und kommutativer Algebra." *Jahresbericht der Deutschen Mathematiker-Vereinigung* 42 pt. 2:132.
title of paper presented to the Mathematical Colloquium of the University Halle-Wittenberg on 30/31 January 1931

1934. *Zerfallende verschränkte Produkte und ihre Maximalordnungen*. Actualités scientifiques et industrielles 148. Paris: Herman & Cie. 15 pp.
dated August 1932

1937. (ed., with J. Cavailles). *Briefwechsel Cantor-Dedekind*. Actualités scientifiques et industrielles 518. Paris: Herman & Cie.

1950. "Idealdifferentiation und Differente." *Journal für die reine und angewandte Mathematik* 188:1-21.
paper presented at the Prague DMV meeting in 1929

II. Reviews and Summaries of Works by Others

1923
In: *Jahrbuch über die Fortschritte der Mathematik* 46

R1. p. 156: Seelig, R. "Über das vollständige Invariantensystem dreier ternärer quadratischer Formen." *Monatshefte für Mathematik* 29 (1918):255–267.

R2. p. 168: Schmeidler, W. "Zur Theorie der primären Punktmoduln." *Mathematische Annalen* 79 (1918):56–75.

R3. p. 671: Loewy, A. "Die Begleitmatrix eines linearen homogenen Differentialausdruckes." *Nachrichten von der Gesellschaft der Wissenschaften zu Göttingen* 1917:255–263.

R4. p. 671: Loewy, A. "Über die Zerlegungen eines linearen homogenen Differentialausdruckes in grösste vollständig reduzible Faktoren." *Sitzungsberichte der Heidelberger Akademie* 1917. 8. Abh. 20 pp.

1924
In: *Jahrbuch über die Fortschritte der Mathematik* 47

R1. p. 85: Hazlett, O. C. "A Theorem on Modular Covariants." *Transactions of the American Mathematical Society* 21 (1920):247–254.

R2. p. 88: Luckhaub, J. "Beiträge zur Geometrie der quadratischen und Hermiteschen Formen." *Monatshefte für Mathematik* 30 (1920):49–58.

R3. p. 88: Coolidge, J. L. "The Geometry of Hermitian Forms." *Transactions of the American Mathematical Society* 21 (1920):44–51.

R4. pp. 88–89: Hurwitz, A. "Über die algebraische Darstellung der Normgebilde." *Mathematische Annalen* 79 (1919):313–320.

R5. pp. 89: Schmeidler, W. "Über die Singularitäten algebraischer Gebilde." *Mathematische Annalen* 81 (1920):223–224.

R6. p. 89: Phillips, H. B. "Functions of Matrices." *American Journal of Mathematics* 41 (1919):266–278.

R7. p. 90: Kostka, C. "Symmetrische Funktionen in Verbindung mit Determinanten." *Leopoldina Nova Acta* 104 no. 3 (1919).

R8. p. 96: Schmeidler, W. "Über Moduln und Gruppen hyperkomplexer Grössen." *Mathematische Zeitschrift* 3(1919):29–42.

R9. p. 97: Schmeidler, W. "Über die Zerlegung der Gruppe der Restklassen eines endlichen Moduls." *Mathematische Zeitschrift* 5 (1919):222–267.

R10. p. 97: Schmeidler, W. "Bemerkungen zur Theorie der abzählbaren Abelschen Gruppen." *Mathematische Zeitschrift* 6 (1920):274–280.

R11. p. 98: Chatelet, A. "Sur les nombres hypercomplexes a multiplication associative et commutative." *Comptes rendus* 169 (1919): 708–711.

R12. p. 133: Brandt, H. "Komposition der binären quadratischen Formen relativ einer Grundform." *Journal für die reine und angewandte Mathematik* 150 (1919–20): 1–46.

R13. p. 134: Gmeiner, I. A. "Über die reduzierten binären quadratischen Formen mit positiver nichtquadratischer Determinante." *Wien. Anzeigen* 56: 195 and "Über die Ketten der reduzierten binären quadratischen Formen." *Wiener Berichte* 129 (1920): 91–127.

R14. p. 146: Bauer, M. "Bemerkungen über die Differente des algebraischen Zahlkörpers." *Mathematische Annalen* 79 (1919): 321–322.

R15. p. 146: Bauer, M. "Zur Theorie der Fundamentalgleichung." *Journal für die reine und angewandte Mathematik* 149 (1919): 89–96.

R16. p. 352: Hensel, K. "Neue Begründung der arithmetischen Theorie der algebraischen Funktionen einer Variablen." *Mathematische Zeitschrift* 5 (1919): 118–131.

R17. p. 353: Hensel, K. "Über die Invarianten algebraischer Körper." *Journal für die reine und angewandte Mathematik* 149 (1919): 125–146.

1925

In: *Jahresbericht der Deutschen Mathematiker-Vereinigung* 33 pt. 2

R1. pp. 151–2: Weitzenböck. *Invariantentheorie*. Groningen, 1923.

In: *Jahrbuch über die Fortschritte der Mathematik* 48

R2. p. 15: Hilbert, D. "Adolf Hurwitz." *Mathematische Annalen* 83 (1921): 161–172.

R3. p. 77: Loewy, A. "Über die Reduktion algebraischer Gleichungen durch Adjunktion insbesondere reeller Radikale." *Mathematische Zeitschrift* 15 (1922): 261–273.

R4. p. 79: Ritt, J. F. "Prime and Composite Polynomials." *Transactions of the American Mathematical Society* 23 (1922): 51–66.

R5. p. 82: Szegö, G. "Bemerkungen zu einem Satz von J. H. Grace über die Wurzeln algebraischer Gleichungen." *Mathematische Zeitschrift* 13 (1922): 28–55.

R6. pp. 88–9: Kempner, A. J. "Über die Separation komplexer Wurzeln algebraischer Gleichungen." *Mathematische Annalen* 85 (1922): 49–59.

R7. p. 93: Mehmke, R. "Einige Sätze über Matrizen." *Journal für die reine und angewandte Mathematik* 152 (1922): 33–39.

R8. p. 104: Hazlett, O. C. "New Proofs of Certain Finiteness Theorems in the Theory of Modular Covariants." *Transactions of the American Mathematical Society* 22 (1921): 144–157.

R9. p. 108: Schmeidler, W. "Über die Singularitäten algebraischer Gebilde." 2. Abh. *Mathematische Annalen* 84 (1921): 303–320.

R10. p. 108: Macaulay, F. S. "Note on the Resultant of a Number of Polynomials of the Same Degree." *Proceedings London Mathematical Society* (2)21 (1922): 14–21.

R11. p. 121: Schur, I. "Über Ringbereiche im Gebiete der ganzzahligen linearen Substitutionen." *Berliner Berichte* 1922: 145–168.

R12. p. 121: Schur, I. "Zur Arithmetik der Potenzreihen mit ganzzahligen Koeffizienten." *Mathematische Zeitschrift* 12 (1922): 95–113.

R13. pp. 195–6: Fujiwara, M. "Anwendung der Geometrie der Zahlen auf indefinite ternäre quadratische Formen." *Jahresbericht der Deutschen Mathematiker-Vereinigung* 30 pt. 2 (1921): 103 and "Zahlengeometrische Untersuchung über die extremen Formen für die indefiniten quadratischen Formen." *Mathematische Annalen* 85 (1922): 21–25.

R14. p. 249: Sternberg, W. "Über die lineare Abhängigkeit von Funktionen mehrerer Variablen." *Mathematische Zeitschrift* 14 (1922): 169–179.

1926

In: *Jahrbuch über die Fortschritte der Mathematik* 47

R1. p. 875: Breuer, S. "Über die irreduktibeln auflösbaren trinomischen Gleichungen fünften Grades." Diss. Frankfurt a.M., Borna-Leipzig: Noske, 1918.

R2. pp. 875–6: Wäisälä, K. "Über den Hilbertschen Irreduzibilitätssatz." *Öfvers. af finska vet soc. förh.* 59A no. 12 (1916–17). 14pp.

1927
In: *Jahresbericht der Deutschen Mathematiker-Vereinigung* 36

R1. p. 71: Study, E. *Einleitung in die Theorie der Invarianten linearer Transformationen auf Grund der Vektorenrechnung.* Part I. Die Wissenschaft, 71. Braunschweig: Vieweg, 1923.

In: *Jahrbuch über die Fortschritte der Mathematik* 49

R2. pp. 71-2: MacDuffee, C. C. "A Theorem on Covariants of Forms with an Application to Linear Algebras." *Bulletin of the American Mathematical Society* 29 (1923):111 and "On Transformable Systems and Covariants of Algebraic Forms." *Bulletin of the American Mathematical Society* 29 (1923):26-33.

R3. pp. 77-8: Fischer, E. "Modulsysteme und Differentialgleichungen." *Jahresbericht der Deutschen Mathematiker-Vereinigung* 32 (1923):148-155 and *Mathematische Zeitschrift* 18 (1923):230-237.

R4. p. 82: Wedderburn, J. H. M. "Algebraic Fields." *Annals of Mathematics* (2)24 (1923):237-264.

R5. p. 82: Krull, W. "Ein neuer Beweis für die Hauptsätze der allgemeinen Idealtheorie." *Mathematische Annalen* 90 (1923):55-64.

R6. p. 89: Dickson, L. E. "A new Simple Theory of Hypercomplex Integers." *Bulletin of the American Mathematical Society* 29:121; *Journal de mathématiques* (9)2 (1923):281-326.

R7. p. 105: Kürschák, J. "Irreduzible Formen." *Journal für die reine und angewandte Mathematik* 152 (1923):180-191.

R8. p. 105: Ballantine, C. R. "Modular Invariants of a Binary Group with Composite Modulus." *American Journal of Mathematics* 45 (1923): 286-293.

R9. p. 106: Fujiwara, M. "Anwendung der Geometrie der Zahlen auf die indefiniten ternären quadratischen Formen." *Hamburg Math. Abh.* 2 (1923):74-80 and "Zur Theorie der binären indefiniten quadratischen Formen." *Tôhoku Math. J.* 23 (1923):76-89.

R10. p. 128: Nagel, T. "Zur Arithmetik der Polynome." In *5. Kongress der Skandinav. Mathematiker in Helsingfors vom 4. bis 7. Juli 1922*, p. 18. Helsingfors: Akademische Buchhandlung, 1923.

1928
In: *Jahrbuch über die Fortschritte der Mathematik* 49

R1. p. 261: Jung, H. W. E. *Einführung in die Theorie der algebraischen Funktionen einer Veränderlichen.* Berlin: W. de Gruyter & Co., 1923. vi+246pp.

R2. p. 459: Linfield, B. Z. "On Certain Polar Curves With Their Application to the Location of the Roots of the Derivatives of a Rational Function." *Transactions of the American Mathematical Society* 25 (1923):239-258.

1929
In: *Jahrbuch über die Fortschritte der Mathematik* 50

R1. pp. 49-50: Breuer, S. "Zur Bestimmung der metazyklischen Minimalbasis von Primzahlgrad." *Mathematische Annalen* 92 (1924):126-144.

R2. p. 50: Schmeidler, W. "Die Übertragung der Galoisschen Aufgabe auf Gleichungssysteme in mehreren Variablen." *Jahresbericht der Deutschen Mathematiker-Vereinigung* 33 (1924):112-116.

R3. p. 50: Sopman, M. "Kriterium der Irreduzibilität der ganzen Funktionen in einem beliebigen algebraischen Körper." *Charikov, Ann. Sc.* 1 (1924):81-82 (in Russian) and "Ein Kriterium für Irreduzibilität ganzer Funktionen in einem beliebigen algebraischen Körper." *Mathematische Annalen* 91 (1924):60-61.

R4. p. 53: Schmeidler, W. "Zur affinrationalen Geometrie." *Journal für die reine und angewandte Mathematik* 153 (1924):215-227.

R5. p. 54: Richardson, R. G. D. "A New Method in the Equivalence of Pairs of Bilinear Forms." *Transactions of the American Mathematical Society* 26 (1924):451-478.

R6. pp. 54-55: Richardson, R. G. D. "Relative Extrema of Pairs of Quadratic and Hermitian Forms." *Transactions of the American Mathematical Society* 26 (1924):479-494.

R7. p. 71: Schmeidler, W. "Über Körper von algebraischen Funktionen mehrerer Variablen." *Nachrichten der Gesellschaft der Wissenschaften zu Göttingen* 1924:189-197.

R8. p. 73: Sono, M. "On the Reduction of Ideals." *Kyoto Science Memoirs* (A) 7 (1924): 191–204.

R9. p. 95: Wahlin, G. E. "On the Application of the Theory of Ideals to Diophantine Analysis." *Bulletin of the American Mathematical Society* 30 (1924): 140–154.

R10. p. 293: Weyl, H. "Zur Theorie der Darstellung der einfachen kontinuierlichen Gruppen. (Aus einem Schreiben an Herrn I. Schur.)" *Berl. Ber.* 1924: 338–345.

R11. pp. 293–4: Schur, I. "Neue Anwendungen der Integralrechnung auf Probleme der Invariantentheorie." I. Mitt. *Berl. Ber.* 1924: 189–208.

R12. pp. 294–5: Schur, I. "Neue Anwendungen der Integralrechnung auf Probleme der Invariantentheorie. II. Über die Darstellung der Drehungsgruppe durch lineare homogene Substitutionen." *Berl. Ber.* 1924: 297–321.

R13. p. 295: Schur, I. "Neue Anwendungen der Integralrechnung auf Probleme der Invariantentheorie. III. Vereinfachung des Integralkalküls. Realitätsfragen." *Berl. Ber.* 1924: 346–355.

1930

In: *Jahrbuch über die Fortschritte der Mathematik* 53

R1. p. 103: van der Waerden, B. L. "Der Multiplizitätsbegriff der algebraischen Geometrie." *Mathematische Annalen* 97 (1927): 756–774.

R2. p. 103: Kapferer, H. "Axiomatische Begründung des Bezoutschen Satzes." (Beiträge zur Algebra 7). *Sitzungsberichte Heidelberg* 1927. 8. Abh. 33–59.

R3. p. 104: Macaulay, F. S. "Some Properties of Enumeration in the Theory of Modular Systems." *Proceedings London Mathematical Society* (2)26 (1927): 531–555.

R4. pp. 116–7: Grell, H. "Beziehungen zwischen den Idealen verschiedener Ringe." *Mathematische Annalen* 97 (1927): 490–523.

R5. p. 117: Grell, H. "Zur Theorie der Ordnungen in algebraischen Zahl- und Funktionenkörpern." *Mathematische Annalen* 97 (1927): 524–588.

R6. pp. 117–8: Hölzer, R. "Zur Theorie der primären Ringe." *Mathematische Annalen* 96 (1927): 719–735.

R7. p. 131: Bell, E. T. "On the Arithmetic of Abelian Functions." *Proceedings USA Academy* 13 (1927): 754–758.

R8. p. 342: Kapferer, H. "Notwendige und hinreichende Multiplizitätsbedingungen zum Noetherschen Fundamentalsatz der algebraischen Funktionen." (Beiträge zur Algebra 8.) *Sitzungsberichte Heidelberg* 1927. 8. Abh.: 61–82.

1931

In: *Jahresbericht der Deutschen Mathematiker-Vereinigung* 40 pt. 2

R1. pp. 11–12: Kronecker, L. *Werke* 4. Ed. by K. Hensel. Leipzig & Berlin: B. G. Teubner, 1929. vi + 509pp.

In: *Jahresbericht der Deutschen Mathematiker-Vereinigung* 41 pt. 2

R2. p. 17: Kronecker, L. *Werke* 5. Ed. by K. Hensel. Leipzig & Berlin: B. G. Teubner, 1930.

In: *Jahrbuch über die Fortschritte der Mathematik* 51

R3. p. 91–92: Dörge, K. "Zum Hilbertschen Irreduzibilitätssatz." *Mathematische Annalen* 95 (1925): 84–97.

R4. p. 92: Dörge, K. "Über die Seltenheit der reduziblen Polynome und der Normalgleichungen." *Mathematische Annalen* 95 (1925): 247–256.

R5. p. 111: Fischer, E. "Über absolute Irreduzibilität." *Mathematische Annalen* 94 (1925): 163–165.

R6. p. 122: Furtwängler, Ph. "Über Minimalbasen für Körper rationaler Funktionen." *Sitzungsberichte der Akademie der Wissenschaften Wien* 134 (1925): 69–80.

R7. p. 143: Bauer, M. "Zur Theorie der algebraischen Körper." *Acta Szeged* 2 (1925): 69–71.

R8. p. 143: Ore, Ö. "Verallgemeinerung des vorstehenden Satzes von Herrn Bauer." *Acta Szeged* 2 (1925): 72–74.

In: *Jahrbuch über die Fortschritte der Mathematik* 54

R9. p. 140: van der Waerden, B. L. "Neue Begründung der Eliminations- und Resultantentheorie." *Nieuw Archief* 15 (1928): 302–320.

R10. pp. 140–1: van der Waerden, B. L. "Die Alternative bei nichtlinearen Gleichungen." *Nachrichten von der Gesellschaft der Wissenschaften zu Göttingen* 1928: 77–87.

R11. p. 141: van der Waerden, B. L. "Eine Verallgemeinerung des Bézoutschen Theorems." *Mathematische Annalen* 99: 497–541 and correction *Mathematische Annalen* 100 (1928): 752.

R12. p. 155: Krull, W. "Zur Theorie der zweiseitigen Ideale in nichtkommutativen Bereichen." *Mathematische Zeitschrift* 28 (1928): 481–503.

R13. pp. 156–7: Krull, W. "Primidealketten in allgemeinen Ringbereichen." *Sitzungsberichte Heidelberg* 1928 no. 7. 14pp.

R14. p. 160: Schmeidler, W. "Grundlagen einer Theorie der algebraischen Funktionen mehrerer Veränderlichen." *Mathematische Zeitschrift* 28 (1928): 116–141.

R15. p. 164: Bell, E. T. "A Generalization of Circulants." *Proceedings Edinburgh Mathematical Society* (2)1 (1928): 177–181.

R16. p. 412: Abramowicz, K. "Transformation des fonctions automorphes." *Comptes rendus* 187 (1928): 801–803.

1932
In: *Jahresbericht der Deutschen Mathematiker-Vereinigung* 42 pt. 2

R1. pp. 38–39: Kronecker, L. *Werke* 3 pt. 2. Ed. by K. Hensel. Leipzig & Berlin: B. G. Teubner, 1931. 215pp.

1933
In: *Jahresbericht der Deutschen Mathematiker-Vereinigung* 43 pt. 2

R1. pp. 94–95: Rutherford, D. E. *Modular Invariants*. Cambridge Tracts No. 27. Cambridge: 1932. viii + 84pp.

MIX
Papier aus verantwortungsvollen Quellen
Paper from responsible sources
FSC® C105338

If you have any concerns about our products,
you can contact us on
ProductSafety@springernature.com

In case Publisher is established outside the EU,
the EU authorized representative is:
Springer Nature Customer Service Center GmbH
Europaplatz 3, 69115 Heidelberg, Germany

Printed by Libri Plureos GmbH
in Hamburg, Germany